Ajay Kumar Goswami, Irena Kostova
Medicinal and Biological Inorganic Chemistry

Also of Interest

Rare Earth Chemistry
Rainer Pöttgen, Thomas Jüstel and Cristian A. Strassert (Eds.), 2020
ISBN 978-3-11-065360-1, e-ISBN 978-3-11-065492-9

Organoselenium Chemistry
Brindaban C. Ranu and Bubun Banerjee (Eds.), 2020
ISBN 978-3-11-062224-9, e-ISBN 978-3-11-062511-0

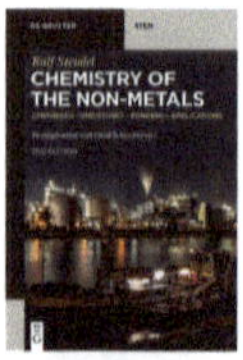

Chemistry of the Non-Metals.
Syntheses - Structures - Bonding - Applications
2nd Edition
Ralf Steudel, in collaboration with David Scheschkewitz, 2020
ISBN 978-3-11-057805-8, e-ISBN 978-3-11-057806-5

Inorganic and Organometallic Polymers
Narendra Pal Singh Chauhan and Narendra Singh Chundawat, 2019
ISBN 978-1-5015-1866-9, e-ISBN 978-1-5015-1460-9

Ajay Kumar Goswami, Irena Kostova

Medicinal and Biological Inorganic Chemistry

DE GRUYTER

Authors
Prof. Ajay Kumar Goswami
132, Road No 5, Subhash Nagar
Udaipur 313001
Rajasthan
India
akumargoswami56@gmail.com

Prof. Irena Kostova
Medical University
Faculty of Pharmacy
Department of Chemistry
Dunav Street 2
1000 Sofia
Bulgaria
irenakostova@yahoo.com

ISBN 978-1-5015-2455-4
e-ISBN (PDF) 978-1-5015-1611-5
e-ISBN (EPUB) 978-1-5015-1615-3

Library of Congress Control Number: 2021950489

Bibliographic information published by the Deutsche Nationalbibliothek
The Deutsche Nationalbibliothek lists this publication in the Deutsche Nationalbibliografie; detailed bibliographic data are available on the Internet at http://dnb.dnb.de.

Cover image: Gettyimages/skynesher
Typesetting: Integra Software Services Pvt. Ltd.
Printing and binding: CPI books GmbH, Leck

www.degruyter.com

Dedicated to our quest for insatiable desire for knowledge.

Preface

Inception of the idea of writing a book on medicinal inorganic chemistry was sown long back while writing a textbook on bioinorganic and supramolecular chemistry. I have always felt that inorganic chemicals and, in particular, metals need a proper place in the medicinal chemistry area and deserve to be equally important compared to their organic counterparts. Bioinorganic chemistry and coordination chemistry were always my research fields, but in this area I was bit hesitant to enter into writing. However, Prof. Kostova's easy consent to write a book jointly made my morale boosted and her contribution as other part was more than encouraging to shape this book. The holistic view of two closely related or rather intertwined branches of emerging field would certainly bring advancement to the research in the area of medicinal and biological inorganic chemistry. This book on alternative system of medicines would hopefully make sensible guidelines for the hitherto advancing medicinal systems and further research on Ayurveda and Tibetan systems in more totality.

The book is a humble approach for merging knowledge in traditional and scientific areas. I leave upon readers to judge and point out shortcomings, if any, and help authors to do still better.

https://doi.org/10.1515/9781501516115-202

Contents

Section A **Medicinal inorganic chemistry**

Chapter 1
Introduction and fundamentals of medicinal inorganic chemistry

Medicinal inorganic chemistry is a young branch of medicinal chemistry emerging due to very promising application of both metals and metal chelates or compounds in medical sciences. The ever-expanding pharmacopoeia of metal-based compounds as therapeutic as well as diagnostic agents further enumerates the significance of this unexplored area of knowledge.

Ever since the evolution of human race, metals have played a vital role. Although it may be a topic of debate whether life and its origin was abiotic or biotic, the significance of inorganic materials including metals is nonetheless most crucial and pivotal in the history of mankind. Even the Indian philosophy describes us as made up of Panch bhutas, the five basic elements of nature. These include air, fire, water, earth and space, and the obvious majority of them include inorganic materials.

Although there is little doubt about significance and contribution of organic reactions in living system both as biological reactions in our body and as medicines, yet innumerable chemical reactions in our body are directly or indirectly dependent on inorganic components.

Synthetic organic chemistry has played a contributory part in the growth of medicinal chemistry, as most of the in vivo reactions of our body are using organic compounds in aqueous media. Nature has been very selective for choosing material for the life processes and thus has chosen metals or their compounds in a justified way. Toxicity has been one major issue, but fairly enough the selectivity does not degrade either of the two organic or inorganic as less important components for the selection but has been with a definite purpose. It is interesting to note that 75% of the chemical elements are metals. The distribution of metals in living systems is under three categories:

(i) The bulk metals such as Na, K, Ca and Mg present as hydrated ions in relatively large quantities, making about 99.5% of the total metal contents associated with the skeleton of vertebrates and osmotic equilibria.
(ii) Essential trace metals, particularly Fe, Zn, Cu, Mn and Mo, are functioning either as metalloenzymes or as enzymatic activators.
(iii) Most of the other remaining metals are placed under this category as they may be accumulated due to specific environmental factors. They are present in tissues or organs and have no biological role per se. Lead and aluminium are such metals which may be present up to 0.03 and 0.06 mg per 100 mL in the whole blood of civilized man with no particular biological function.

https://doi.org/10.1515/9781501516115-001

In essence, metals in biological systems are complementary as well as supplementary and we cannot ignore their role as an ingredient of medicinal value. Metal toxicity has been an issue of prime attention but with precaution the application and research based on cautious but purposeful use of metals is warranted.

All metal ions that have gained access to a biological system in sufficient amount are not rapidly excreted but are toxic depending on the individual metal classified above as bulk, trace or essential. They can partake by either deranging the electrolytic balance, as irritants to damage that specific organ such as kidney or by affecting the central nervous system and by interfering with the enzymatic process. There can be considerable difference in the level of toxic concentration but there is no suggestion that an organism is tolerant to high concentration of metals serving a useful biological purpose.

1.1 Metals in ayurvedic system as medicines

A very comprehensive review based on *Charaka Samhita* has beautifully presented the application of metals in this ancient medicinal pharmacopoeia [2]. The *Charaka Samhita* or *Compendium of Charaka* is a Sanskrit text of Indian traditional medicine. *Charaka Samhita* and *Sushruta* are two foundational texts of this field. This review is based on ancient compendium of traditional materia medica. It is worth noting that during the medieval period *Rasashastra*-based use of certain heavy metals and minerals in ayurvedic therapeutics increased. *Rasashastra* is an integral part of ayurveda dealing with drugs of mineral origin. It describes details of their varieties, characteristics, processing techniques, properties, therapeutic uses, possibilities of developing adverse effects and finally their management. Although roots of this science exists in ancient texts of Indian civilization, its development as an independent system of therapy started around the eighth century AD. Ayurvedic classics written before that time like *Charaka Samhita* and *Sushruta Samhita* contain description of metals and minerals, their processing techniques and their utilization in therapeutics.

1.2 Metal chelates and their application in medicinal chemistry

The second important part of this book is the application of metal complexes and metal chelates in medicines. A very simple approach on this topic was presented in the form of an edited book long back by Dwyer and Mellor [3]. Use of chelating agents in medicine basically depends on the formation of a soluble and easily excretable metal chelate by hiding the metal ions in the circulation or competing the chelating biological site for the bound metal ion. The major conditions for this process are:

(i) The chelating agent must have low toxicity.
(ii) It should not readily metabolize.
(iii) It should be capable of penetrating to the metal storage site. Even, in general, the metal chelate should be less toxic than the free metal ion.

The introductory medicinal inorganic chemistry discussion can be concluded with the following highlights:

- Metals have been used in treatment since ancient times. The *Ebers Papyrus* from 1550 BC is the first written account of the use of metals for treatment and describes the use of Cu and Fe.
- Although dates of *Charaka Samhita* are not certain, Meulenbeld's *A History of Indian Medical Literature* dates it to be between fourth century BCE to second century CE, with *Compendium of Charaka* to be between 100 BCE and 200 CE [4]. This system of ancient times has references of use of metals as medicine. The word 'Bhasma' is the name of metal-based medicines, including copper (Tamra), gold (Swarna) or mercury (Parad) in the palatable form.
- Even Chinese and African systems of medicines use metals or their compounds as medicines.
- Homeopathy incorporates metals or metal-based compounds for treatment.
- Application of metals to medicines is a rapidly developing area, and novel therapeutic and diagnostic metal complexes are now having great impact on medical science.
- Copper was used to sterilize water in Egypt as far back as 3000 BC.
- The Chinese and Arabs used gold in a number of medicines over 3,500 years ago believing that the precious metal possessed medicinal value.
- During Renaissance period in Europe, mercurous chloride (Hg_2Cl_2) was used as a diuretic.
- Paul Ehrlich, the so-called founder of chemotherapy developed the arsenical Salvarsan as a drug for treatment of syphilis in the early twentieth century.
- It is estimated that about 80% of South Africans use some form or other of traditional medicines for their primary health care, which include metals/metal-based compounds (WHO 2010). Heavy metals may enter traditional medicines by means of both intentional and unintentional routes.
- In traditional South African medicinal markets and medicinal shops, a number of powdered synthetic materials known as 'imikhandu' (in ISI Zulu language) are readily available. Such crystalline salts or even metallic compounds are examples of medicinal application of inorganic compounds. Some examples are:
  (i) ammonium chloride,
  (ii) sulphur powder,
  (iii) copper sulphate,
  (iv) potassium dichromate/sodium chloride,

(v) plant resins and
(vi) potassium permanganate.

Conclusively, this book will discuss and describe the underlying principles, drug action, possible mechanism and future prospects of medicinal inorganic chemistry as an emerging branch of knowledge.

Chapter 2
Metals in medicine: perspective and prospective view

2.1 Metal–ligand bonding

To define the structure of a metal complex (a species formed by the association of two or more simpler species each capable of independent existence) [5], we are supposed to know the coordination number of the central metal, stereochemistry, conformation of a molecule or complex ion and finally the nature of the bond between a metal and a ligand. They can all explain the interpretation of the structure and reactivity of a metal complex.

The metal–ligand bond, in general, is described in terms of fundamental atomic properties of electron – configuration, the nature of orbitals involved, overlapping powers and the ionization potentials. The nature of metal–ligand bond, interpretation of properties of complexes and even the geometries or stereochemistry of metal complexes are precisely explained on the basis of theories such as Crystal Field Theory (CFT), Ligand Field Theory (LFT) or Molecular Orbital Theory (MOT). These theories are part of learning of any coordination chemist. This part here is simply a review of nature of metal–ligand bonding. Metal–ligand bonding varies from extreme ionic to covalent type and even within these two ends intermediate types are σ-type, π-type, multicentre as well as ion–ion to ion–dipole type. A brief description can be presented by a linear figure (Figure 2.1).

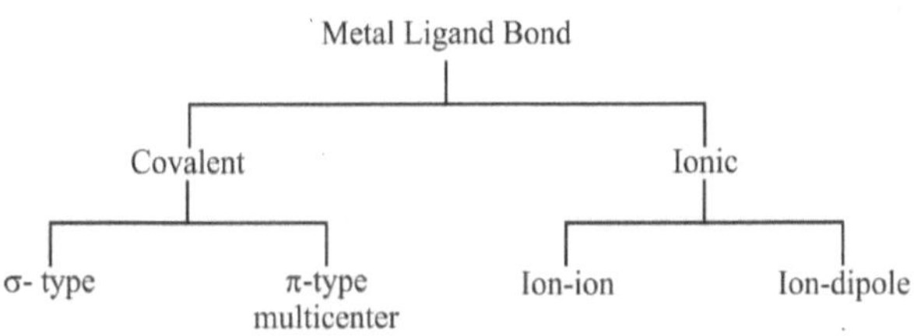

Figure 2.1: Types of metal ligand bonds.

An important aspect of metal–ligand bonding cannot be complete without including transition metals. The transition metals are metals having a d^1–d^9 electronic configuration. The root of coordination behaviour lies in properties such as variable oxidation states, magnetic properties, stereochemistries and even stability constants of the complexes. With increasing oxidation state, covalent characters increase. However, within the same oxidation state with an increase in the number of

https://doi.org/10.1515/9781501516115-002

d-electrons, the covalent character increases. The fundamental properties of the metal complexes or transition metal complexes to be specific are explained by the crystal field model, which is purely an ionic model, whereas the ligand field theory explains the effect of covalent bonding [6–8].

2.2 Bioinorganic chemistry

The application of metals in medicine can be more deeply explored if we understand how metals are associated with our biological functions. A host of literature on this, under bioinorganic chemistry, is available. The bioinorganic chemistry is basically concerned with the function of all metallic and even non-metallic elements in biological systems both in vitro and in vivo. However, recent development of the branch has centred upon

(i) Improved method of preparations
(ii) Synthesis of inorganic complexes and metal chelates as mimics for various naturally occurring biomolecules
(iii) Introduction of metals into biosystem as drugs or even probes
(iv) Application and research on the role of inorganic elements in toxicity and nutrition
(v) Increased awareness about the environmental hazards causing danger to living beings
(vi) Recognition of significance of an increasing number of trace elements in plant, animal and human nutrition

Before launching the main theme of this book, that is, how metals can be futuristic drugs, a brief description of classification of metals and their association with specific biological function is warranted (Table 2.1). As we know that the elements of biological importance associated with functionality have been classified on the basis of ash of human tissue, it is found that at least 30 elements in different categories are vital. The three categories are:

(i) **Essential elements**: These elements are indispensable for maintaining normal living stage of a particular tissue or whole of the body. Depending on their absolute amount in the body, they have been further divided into two subgroups: (a) macro-essential elements (macro-nutrients), (b) micro-essential elements (micro-nutrients)
    (a) ***Macro-essential elements:*** The elements that are required to be present in the diet in quantities of more than 1 mg are called macro-essential elements. They form nearly 60–80% of all inorganic matter in the body. These include elements such as C, H, O, N, Na, K, Ca, Mg, Fe, P, S and Cl. Out of these, the first four elements occur in substantial amounts in energy tissue and get derived from dietary carbohydrates, lipids and proteins. Oxygen is

obtained directly from atmosphere. Nearly 85% of total oxygen and 70% of total hydrogen are available in the form of water, making three-fifth of the total body weight. The remaining amount of O, H, all N, C, S and others are derived from carbohydrates, lipids and proteins which help support the basic requirements of tissues and synthesis of various biochemical substances within these structures. The elements that are required to be present in the diet in quantities of less than 1 mg are called micro-essential elements. These include Cu, Zn, Co, Mn, Mo, I and F.

(ii) ***Non-essential elements***: Although they are present in the body but are not essentially associated with a known function, they might be termed as non-essential. These elements are B, Si, As, Li, Al and Pb. A schematic diagram (Figure 2.2) summarizes the description.

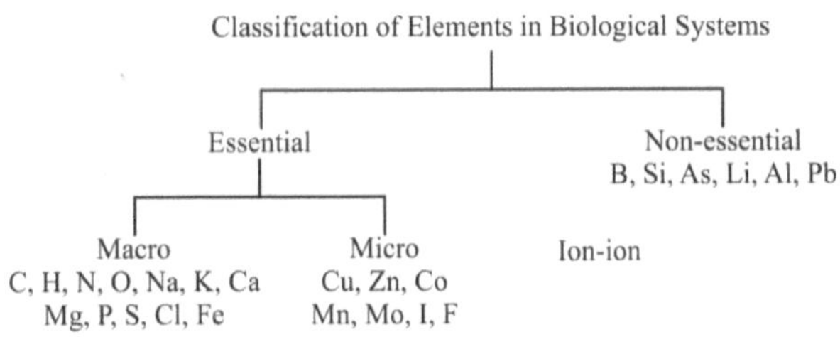

Figure 2.2: Various elements in biological systems.

Table 2.1: Classification based on the function of complex biological systems.

Elements in extracellular fluid	Elements inside the cell	Elements in cytoplasm
Na, Ca	K, Mg	K, Mg
Cu, Mo	Fe, Ca	Ca
Cl, Si	Zn, Al, Mn	Zn
	P, S	P, S
	Se	Se

Specificity is a buzzword in selection of individual elements by nature during evolution. There are elements with limited biological function which are present in biological systems. On this concept, it is assumed that a peculiarity is the result of evolutionary drive toward the best chemical system for a living organism. Table 2.2 gives a brief classification on this basis, and the elements have been classified as per their quantities in the system.

Table 2.2: Essential elements.

Bulk structural elements	Mineral elements	Trace elements	Ultra trace	Non-metals
H	Na	Fe	Mn	F
C	K	Cu	Mo	I
H	Mg	Zn	Cr	B
O	Ca	V	Si	
P	Cl	Ni	As	
S		Cd, Li, Pb, Sn	Se	

The importance of metals is explained by the fact that both deficiency as well as excess of metals is alarming in biological system. A group-wise classification can further signify their importance to conclude, and Table 2.3 gives this description:

Table 2.3: Group-wise biologically important elements.

Group IA	Group IIA	Group IIB	Transition metals
Na^+, K^+	Mg^{2+}, Ca^{2+}	Zn^{2+}, Ni^{2+}	Mn, Fe, Co, Mo
Useful osmotic equilibrium	Triggers conformational control	Acid catalysis	Redox catalysis

2.3 Importance of metal-based drugs in medicinal inorganic chemistry

There have been mentions about historically proven use of metal-based remedies by ancient civilization of Mesopotamia, India, Egypt, China and Africa. Metallodrugs have been in use for minor stomach pains or indigestion for centuries. We know how cisplatin has offered a drug for cancer, and metals are not less important in diagnosis such as in MRI. The metal-based drugs or medicinal inorganic chemistry is thus a branch which is now holding a promising future. After the approval of cisplatin (*cis*-diamminedichloroplatinum) as a chemotherapeutic for cancer by FDA in 1978, the focus of medicinal chemists has been drawn to develop efficient metal complexes for their application in medicine. Metallodrugs have found successful applications in cure or treatment of diabetes, ulcers, inflammation, arthritis and even cardiovascular diseases. Although alternative systems of medicine such as Ayurveda or homeopathy, Chinese or African systems too have significant use of metals or metallodrugs, the mechanism of drug action on the principles of scientific drug discovery steps is in its age of development. This book incorporates in brief the mechanism of drug action and the recent state of the art of various metals which have been used in the treatment of a particular disease.

2.4 Biological assays

Any therapeutic agent or drug needs to be tested in tissue culture on a suitable model system of late theoretical prediction of activities using computer software such as PASS (prediction of activity structure spectra), molecular docking and quantitative structure–activity relationship are frequently being used for any new lead to be tested a priori. This not only saves time but also gives scope for a better strategy not involving direct synthesis and then getting no-activity or not the desired one. Thus, exploration of new lead and its exploitation are the two mainstays of medicinal chemistry [1, 2]. Excellent reviews on theoretical aspects of drug design and development are available in literature [3, 4].

Application of any inorganic compound or metallic compound to medicine warrants a detailed examination of the fundamental aqueous chemistry of the compound. The examination includes its pharmacokinetics, metabolic fate in intracellular blood and its effects on the target of choice. The issue of toxicology of inorganic compounds in medicinal use, particularly those with heavy metals, has a kind of stigma, degrading the object of drug development. Although the therapeutic softwares are well defined to minimize side effects and usefulness, any drug keeping a tightrope balances between activity and toxicity.

2.5 Metal ion toxicity

As described in the earlier section of introduction, the distribution of about 75% of chemical elements (metals) is categorized under three heads: bulk, essential and non-essential elements. Their action in biological system is also well understood. Many soluble metal salts if ingested can produce local irritation and tissue damage. Systemic poisoning may also occur if the metal ion is absorbed in sufficient quantity and gets into the circulation. Emesis is caused due to absorption of Zn and copper salts including irritation of gastrointestinal tract. Since absorption is low, there are seldom chances of systemic poisoning. It is very common knowledge that Cu deficiency causes Wilson's disease. This is due to deranged copper level control which causes increased absorption to the extent that excretion cannot dispose of the excess. This results in accumulation in the liver.

In general, absorption of metal ions through gastrointestinal tract is poor. Although Na^+, K^+ and Ca^{2+} are exceptions, a number of organic cations are also in the same way poorly absorbed. Even complex cations are poorly absorbed or even not absorbed. Since Hg^{2+} ion quickly forms a neutral molecule $HgCl_2$, it is rapidly absorbed. The mechanism of increased uptake of copper in Wilson's disease is yet not sure as if it is through coordination. With iron, the absorption is studied in detail; it is through duodenum and jejunum where it is captured by the mucosal cells by protein apoferritin. It can be further understood that in the form of ferritin which is hydroxoiron(III) phosphate or a polymeric diol-bridged complex, iron is in equilibrium with the ferrous ion of the mucosal cells and the plasma iron of the blood stream [5, 6].

The toxicity of any inorganic compound primarily may derive from its metabolism and undesirable interaction with random proteins. The better approach thus can understand the route and site of interaction. The mechanistic approach simplifies the understanding of interaction and toxicity, thus enabling us to design our drug on this basis.

2.6 Drug interaction

An excellent review on metal complexes as drugs and chemotherapeutic agents addresses fundamental issues and mechanism of drug interaction in case of metals [7]. Metal ions, or drugs and metabolite, including endogenous ligands are transported through human serum albumin (HSA) [8]. HSA is one of the most common proteins in the circulatory system that plays a significant role in the transport of metabolites and drugs [9]. This transport protein plays a vital role in protein binding for many drugs. The HSA has a very simple structure representing a quaternary model. The molecular mass of the protein is 67,000 with 585 amino acids [10]. HSA can be represented by the structure shown in Figure 2.3.

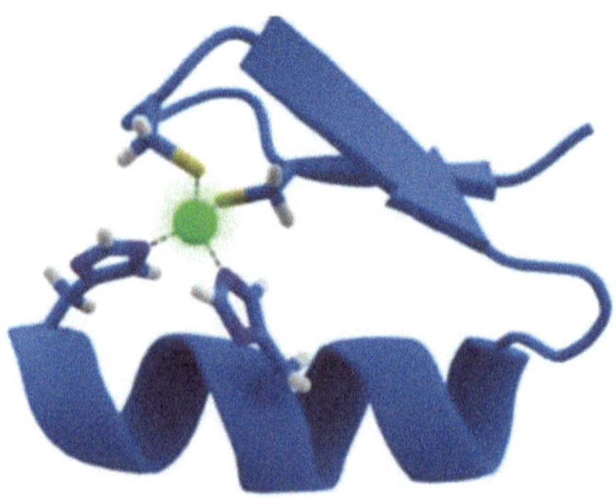

Figure 2.3: Zinc Finger.

The understanding of the binding sites can be visualized by the crystal structure and physical and biophysical studies for many inorganic compounds. Literature reveals the abundant cysteine-rich small peptides, particularly glutathione (GSH). Sulphur-containing metal-binding proteins are shown in Figure 2.4(a) and (b).

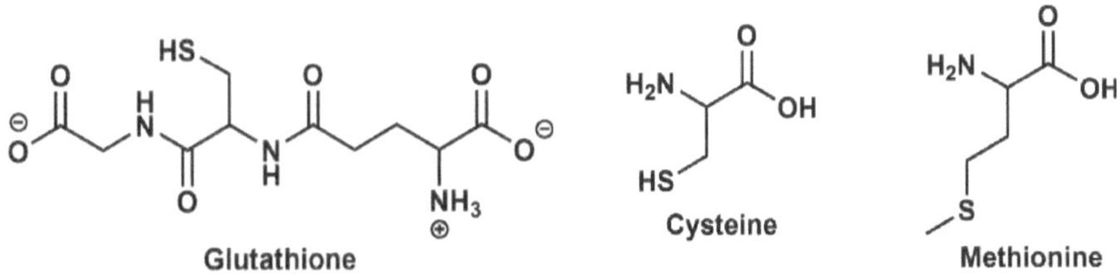

Figure 2.4: (a) Glutathione and (b) cysteine and methionine.

Structure of other such proteins is presented in Figure 2.4. Proteins like metallothionein represent both detoxifying and deactivating pathways for inorganic drugs. Transferrin is another protein involved in the protein–metal ion interaction providing natural iron-binding site which is also accessible to the number of metal ions including Ru^{2+}, Ga^{3+} and Al^{3+}, owing to similar charge-to-radius ratio. Thus, the nature of target to be attacked by any drug depends on its specific application. A number of cytotoxic metal complexes target DNA due to its importance in replication and cell viability. In general, coordination compounds use many binding modes to polynucleotides incorporating outer sphere non-covalent bonding of metal coordination to nucleo-based and phosphate backbone site. They also involve strand cleavage induced by oxidation through redox-active metal centres. Such purine and pyrimidine mononucleotide building blocks are presented in Figure 2.5.

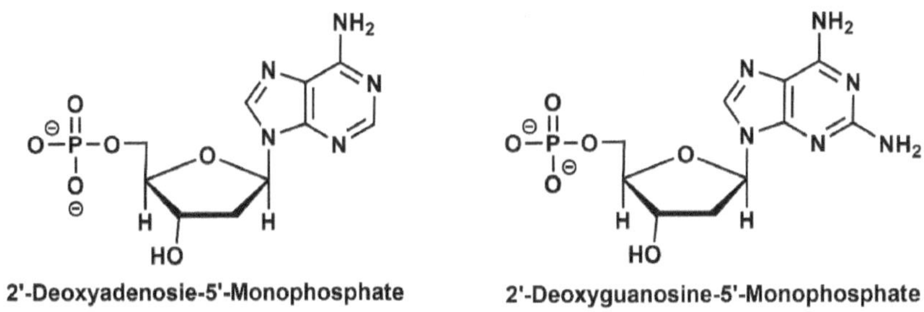

Figure 2.5: Purine and pyrimidine mononucleotide building blocks.

Structure of nucleic acid monophosphates, metal-binding centres (Farrel) and other transition metals like platinum and ruthenium are favourably binding to electron-rich nitrogen, particularly guanine N7. Titanium and early metals display a mixture of nucleobase and phosphate backbone bonding.

Although it is difficult to give a detailed description of metal–nucleic acid interaction, here a general description is being given to present a simple mechanistic view.

A very precise detail of metal–nucleic acid interaction is needed to explain the drug interaction and toxicity particularly related to heavy metals. A point-wise highlight of different types of metal–nucleic acid interaction can be classified as follows:
(i) Coordination
(ii) Intercalation and hydrogen bonding
(iii) Fundamental reaction with nucleic acids
(iv) Structural role of metal–nucleic acid interactions
(v) Regulatory role of metal–nucleic acid interactions
(vi) Pharmaceutical role of metal–nucleic acid interaction

(i) **Coordination**: One of the most prevalent interactions of metals with nucleic acids is through coordination. The soft metal ions are coordinated at the nucleophilic position of the bases; for example in *cis*-$(NH_3)_2$ Pt-dGpG, the platinum centre coordinates to the N_7 position on cytosine, the deprotonated N_3 position on thymine and uracil and N_1 position of purines have also been observed, and even metal ions with decreasing softness have been observed to be potent for coordinating to the phosphate O_2 atoms.

   The metal ions involved, however, decide on the ionic versus covalent nature of these complexes. The DNA–helix melting temperature is the tool to establish metal ion preference for base versus phosphate binding. A tentative order can be expressed by Mg(II) > Co(II) > Ni(II) > Mn(II) > Zn(II) > Cd(II) > Cu(II). However, it is observed that the pentose ring is a poor ligand for coordinating with metal ions.
(ii) **Intercalation and hydrogen bonding**: The DNA metal or metal complex interaction is generally a non-covalent interaction such as
   (a) metallo-intercalation and
   (b) hydrogen bonding interaction.

   Planar aromatic hydrocyclic ligands, for example phenanthroline and tetrapyridine, stack between the DNA base pairs through dipole–dipole interaction. A mixed type of non-covalent–covalent interaction may also occur.
(iii) **Fundamental reaction with nucleic acids**: These reactions, in general, fall under two major categories:
   (a) those involving a redox reaction mediating oxidation of nucleic acid and
   (b) those involving coordination of a metal centre to sugar phosphate backbone mediating hydrolysis and polymer.

   The simplest polynucleotide reaction with metal via redox reaction mechanism can be exemplified by the Fenton reaction which indirectly promotes DNA strand scission through radical reactions on sugar ring:

$$Fe(II) + H_2O_2 \rightarrow Fe(II) + O\dot{H} + OH^-$$

$$Fe(III) + e^- \rightarrow Fe(II)$$

The hydroxyl radicals produced by the Fenton reaction with hydrogen peroxide mediate the cleavage of DNA strand by coordinating to the sugar phosphate backbone.

(iv) **Structural role of metal–nucleic acid interactions:** One of the major functions attributed to metal ions in any biological system is their capacity to facilitate the centre of a structure that directs folding of a protein. This would be apt to mention here that the metalloproteins that have received great attention of late are 'zinc-finger' regulatory proteins. They are relatively small protein motifs containing multiple finger-like protrusions which make tandem contacts with their target molecules. The term zinc finger was coined to describe the finger-like appearance of a hypothesized structure from *Xenopus laevis* transcription factor IIIA (TFIIIA), which is how used to describe a wide variety of differing protein structures.

The zinc ions play an important role in the functioning of the nucleic acid binding TFIIIA, specifically DNA and RNA.

(v) **Regulatory role of metal–nucleic acid interactions:** Changing intracellular metal concentration and the response to it by a biological system is an interesting stimuli–response phenomenon. At a high concentration, many metals become toxic to the cell, and thus need a metalloregulatory protein which binds DNA in the absence of metal ions and represses transcription.

In the presence of metal ions, these regulatory proteins bind metal ions tightly and specifically resulting in amplification of transcription. Best example of this is MerR system which regulates the mercury in bacteria. Fe(II) and Cu(II) binding systems in yeast are other such examples.

(vi) **Pharmaceutical role of metal–nucleic acid interaction:** The intricate understanding of metals as medicines can be understood by interpreting the major pharmaceutical role of interaction of metals with the nucleic acids.

Most pharmaceutical compounds being used are DNA-binding agents including peptide and/or saccharide functionalities. They bind DNA by either of the three routes through intercalation, groove binding or a combination of both. Metal ions, however, appear to be necessary to the functioning of many enzymes acting on nucleic acids; examples are Zn ions, which are essential to the functioning of both RNA polymerase and DNA topoisomerases. DNase I also requires Ca^{2+} for its catalytic activity. Eco RI* (Eco R one) too requires Zn^{2+} and DNA repair enzyme. Further, endonuclease III also needs Zn^{2+} for its catalytic activity.

Eco RI is a restriction endonuclease enzyme isolated from species *E. coli*. The Eco part of the name originates from the species and R represents the particular strain RY13, and 1 describes that it was the first enzyme isolated from this strain.

2.6.1 Metals as medicines: a review

Although metal-based drugs including imaging agents have been in use for the clinical treatment and diagnosis of cancers, a variety of other diseases have also been treated using such compounds. As emphasized earlier, the advantage of these compounds is their possibility of using alternative mechanism of action as compared to other drugs.

Before presenting a brief review on medicinal use of individual metal or even non-metal (as inorganic medicine), there is a need to describe in short role of some such elements. Table 2.4 summarizes this role in terms of disease arising and toxicity, if any.

Table 2.4: Disease associated with low and high level of certain elements.

S. no.	Element	Disease caused due to deficiency	Disease associated with excess
1	Calcium	Bone deformities	Cataracts, gallstones, atherosclerosis
2	Cobalt	Anaemia	Coronary failure polycythaemia
3	Copper	Anaemia, kinky hair syndrome	Wilson's disease
4	Chromium	Incorrect glucose metabolism	Primary irritant dermatitis
5	Iron	Anaemia	Liver disease, heart problem and diabetes
6	Lithium	Manic depression	Tremor, kidney problem, altered level of consciousness
7	Magnesium	Convulsions	Hypermagnesemia, renal failure
8	Manganese	Skeletal deformation, gonadal dysfunction	Ataxia
9	Potassium	Weakness, fatigue	Kidney disease, Addison's disease
10	Selenium	Necrosis of liver, white muscle disease	Blind staggers in cattle
11	Sodium	Addison's disease, stoker's cramps	Risk of stroke, heart failure, kidney disease, osteoporosis
12	Zinc	Dwarfism, hypogonadism	Metal fume fever
13	Cadmium		Nephritis, anaemia
14	Lead		Encephalitis, neuritis
15	Mercury		Encephalitis, neuritis

It is interesting to note that many other alternative systems of medicine like African, Chinese Ayurvedic, Tibetan Ayurvedic and even homeopathy include inorganic medicines as part of their pharmacopoeia. Since it needs a lot of research before any standard drug is marketed, this is a grey area of research. The book simply includes a brief description of this hitherto underplayed area, which needs focus. To be more precise, a metal-wise review of application as a medicinal compound is being presented. Table 2.5 depicts the list of inorganic compounds as medicine.

Table 2.5: Examples of metal-based drugs.

	Metal	Product name	Active compound	Medicinal use
1	Li	Camcolit	Li_2CO_3	Manic depression
2	Mg	Magnesia	MgO	Laxative
3	Fe	Fentamine	$Na_4Fe(II)$ citrate	Anaemia
4	Co	Cobalamin S	Vitamin B_{12}	Anaemia
5	Zn	Calamine	ZnO	Skin ointment
6	Ba	Baridol	$BaSO_4$	X-ray contrast medium
7	Pt	Cisplatin	*cis*-$[Pt(NH_3)_2Cl_2]$	Anticancer
8	Au	Auranofin	Au(I) (PEt_3) acetylthioglucose	Rheumatoid arthritis
9	Bi	De-Nol	$Ke[Bi(iii)(citrate)_2]$	Antiulcer

On the basis of review, it is observed that apart from Table 2.5, the inorganic compounds, particularly metal-based compounds have been used in curing or diagnosing diseases. A list of diseases that uses inorganic compounds will justify the observation signifying application. The compounds have been used as:

(i) Antidepressants
(ii) Insulin mimetic
(iii) Diagnostic agents, X-ray/MRI
(iv) Radiopharmaceutical
(v) Anti-cancer
(vi) Anti-arthritic
(vii) Anti-infective
(viii) Anti-ulcer

Not only these diseases, a list of other medicinal application of metal-based drugs is also found in the literature. Table 2.6 gives a brief description of disease-wise medicinal application of such metal-based drugs.

2.6.2 Use trade names/activity compounds

Table 2.6: Disease-wise application of metal-based drugs.

S. no.	Action	Drug	Compound	Metal
1	Anti-bacterial disinfectant	Thiomersal mercurochrome	Hg-organic compound $AgNO_3$ Ag (sulphadiazene)	Hg Ag
2	Anti-fungal	Thiomersal mercurochrome	Hg-organic compound $AgNO_3$ Ag (sulphadiazene)	
3	Anti-cancer	Cisplatin satraplatin AMD-473 SoD mimics NAmi-A Trisenox, use in acute promyelocytic leukaemia	Polynuclear Pt (IV) species Mn chelates Trans-$[RuCl_4(Me_2SO_4)$ $*IM]^-$ As_2O_3	Pt Mn Ru As
4	Antacid anti-ulcer	Pepto-Bismol ranitidine bismuth De-Nol	Bi(sugar) polymers	Bi
5	Vasodilator	Nipride for acute shock No release	$[Fe(No)CCN)_5]^{2-}$	
6	Hypercalcaemia of malignancy	Ganite possible anti-cancer agent	$Ga(NO_3)_3$	Ga
7	Type II diabetes	BMOV, insulin mimetic	$VO(maltate)_2$	V
8	Phosphate binder fosrenol	Hyperphosphataemia	$Ln(CO_3)_3$	Ln
9	Rheumatoid arthritis	Orally active Ridaura	$Au(PEt_3)$ acetylthioglucose	Au
10	Anti-parasitic leishmaniasis	Sub(III) (tartrate) Tartar emetic stibophen, astiban		Sb

2.6.3 Review of individual metallodrugs

2.6.3.1 Medicinal application of silver (Ag)

A number of reviews on historical use of silver in medicinal application are available in the literature [11]. It is reported that ionized form of silver (Ag^+) possesses known antibacterial properties. There are a number of wound dressings containing silver that are marketed. Even a number of consumer products like clothing and household appliances are being promoted of late. Even historically silver has been used for medical applications. A history of medicinal use of silver further warrants research on its use [12]. It is worth mentioning that it is effective against almost all organisms tested and even has been used for the treatment of number of infections with surprising success. Even its radiological application is worth mentioning.

2.6.4 Absorption and metabolism of silver

A review on pharmacological and toxicology of silver describes the absorption and metabolism of silver [13]. The review mentions that silver is absorbed in the human body through ingestion, inhalation and intraparenteral insertion (by medical device).

Silver being a highly reactive metal is absorbed by either of the routes, such as ingestion, inhalation and percutaneous absorption. It is metabolized in our body as a complex [14] by use of metal carrier proteins, which are deposited in bones and soft tissues. The metallothioneins are induced by silver and are the major carrier proteins. These are cytoprotective proteins. However, macrophage sequestrating silver acts as an additional cytoprotective mechanism for the body. It is not regarded as a cumulative toxin and excreted through the urinary tract. Although minimal toxicological significance is attached to silver, there may be a small residual portion remaining to attach to hydroxyapatite in the bones to silver, yet the principal site for its deposition is called argyria discoloration, which is shown dermal regions of the skin.

2.6.5 Medicinal application of gold (Au)

Biomedical application of gold – a brief history published by the World Gold Council mentions that gold is becoming extremely important in medical treatment ranging from its use as drugs to precession implants.

Further, a review of historical use of gold in medicine [15] presents an overall view on how long the gold has remained as part of our medicinal system. Ever since the development of human civilization, gold compounds or gold were used both by Sharman and medical practitioners for the treatment of diseases or pathological problems. Even injectable gold compounds have been successfully used for the treatment of rheumatoid arthritis. However, out of the number of gold compounds, gold sodium thiomalate is one of the most extensively researched compounds by both clinicians and scientists. A drug called auranofin which showed promising therapeutic potential was also used. This drug offered competition to the injectable gold. The only problems which caused doubt against this drug were clinical side effects and fear of immune suppression by the drug. Thus, gold sodium thiomalate still continues as the preferred gold compound for the management of rheumatoid arthritis. Further, increased use of low-dose methotrexate (MTx) is also an optional drug being used. There is no doubt that injectable gold compounds are indeed a first choice for treatment of the rheumatoid arthritis till data.

2.6.6 Medicinal application of arsenic

The relation between toxic action and metal coordination was proposed by Voegtlin et al. [16]. It was advanced by them that toxic action of the metalloid arsenic on living cells is due to its combination present in the protoplasm. However, in recent years, arsenic-based compounds have been accepted as medicine for cancer therapy as a result of their high rate of success, particularly some cancers like acute promyelocytic leukaemia (APL) [17].

The mechanism of arsenic-containing compound has now been portrayed or precisely understood. The mechanism by which these compounds kill any particular types of cells is selective killing of cancer cells. The information was gained parallel to understanding of increased importance of intracellular redox systems and regulation of the production of reactive oxygen species (ROS) through mitochondrial function.

Many of these targets for arsenic-containing compounds are mitochondrial proteins which are associated with the production of ROS. So, by the disulphide linkage of vicinal thiol groups often leads to increased production of ROS which inducts apoptotic signalling pathways. The details of the mechanism are beyond the scope of this book. However, arsenic compounds with modified organic structures have found a place as anti-neoplastic agents. The compounds have been synthesized and examined for their efficacy against human leukaemia cells and even breast cancer cells in culture. Some of these compounds have exhibited cytotoxic anticancer activity against human breast cancer cells. A novel glutathionyl peptide trivalent arsenic containing compound *para*-4-[*N*-(*S*-glutathionyl acetyl)amino] phenyl arsenoxide has shown potential as an anti-neoplastic drug. As(III) as the anhydrous form of $AS(OH)_3$ (Trisenox, Cell Therapeutics, Seattle, WA, USA) received FDA approval in 2000 as a chemotherapeutic agent for the treatment of APL [19].

2.6.7 Medicinal application of aluminium

A number of drugs containing aluminium hydroxide are used to treat heartburn and stomach upset such as antacid or acid-indigestion conditions: Amphogel, Alucap, Dialume Alu-Tab, Alternagel or Aloh-Gel are some of the brands under which the drugs are sold. It is also used to reduce phosphate levels in people with certain kidney condition. These compounds have been used as phosphate binders and as adjuvants in vaccine. Aluminium is also part of implant materials, for example glass ionomer cement. Aluminium acetate is an astringent.

2.6.8 Medicinal application of bismuth

The commonest compound sold as generic under brand name Pepto-Bismol is bismuth subsalicylate. The drug is used to treat temporary discomfort of the stomach and gastrointestinal tract. Management to diarrhoea, indigestion, heartburn and nausea is done using the compound also known as 'pink bismuth'. The drug is used for prevention and treatment of traveller's diarrhoea. A review of medicinal use of bismuth compounds is available in the literature [20]. Although bismuth-based drugs are primarily used for the treatment of gastrointestinal disorders, it also has significant anti-microbial, anti-leishmanial and anti-cancer properties [21]. Owing to this, lot of research has been done on synthesis, characterization and bioscreening of new bismuth-based compounds. Their potential as anti-microbial, anti-leishmanial and anti-cancer agents has been thoroughly evaluated. As described under section of drug action, it is probable that proteins are the key biomolecular targets of bismuth and particularly thiol-rich ones like cysteine offer attractive metal coordinating sites. It is worth mentioning that metallomic/metalloproteomic understanding approach can help in explaining the mechanism of action of Bi on the biomolecules. It will further the elucidation of mechanism of resistance employed by protozoan, microbial or even carcinomal cells. The strategy will explain the mechanism of uptake, storage and removal of Bi [22].

^{212}Bi- and ^{213}Bi-containing radio-pharmaceuticals are promising effective α-emitting anticancer agents. The radiolabelling using suitable bidentate chelating ligands with both ^{212}Bi and ^{213}Bi and selecting effective targets can be a right plan of research in future. However, this is certain that owing to very great potential, Bi can play a very significant role as a medicinal candidate.

2.6.9 Medical applications of beryllium

Beryllium has a low atomic mass and density and is transparent to X-rays and other ionizing radiations. This has made it a key component in the making of X-ray windows. There are some silver applications of beryllium in pacemakers. CAT scanners, MRI machines, laser scalpels and even in springs and membranes for surgical instruments as beryllium iron or beryllium nickel alloys. The toxicity of this element restrains its use directly as a compound for medicine. It may be noted that acute exposure to beryllium fumes can cause severe pneumonitis. Its poisoning is termed as berylliosis.

2.6.10 Medicinal/medical applications of barium

The commonest use of barium or its compounds is in the medical procedure. It is called barium enema when doctors need to examine the digestive system of any patient. A mixture containing barium sulphate is used to coat the inner lining of intestines. Further, if the stomach and oesophagus are examined, then the patient is administered chalky barium sulphate liquid. When the patient is x-rayed, this coating of barium sulphate absorbs major proportion of x-ray radiations. This results in black and white contrast to facilitate the doctor better diagnose the digestive problem. Barium sulphate is thus called a contrast agent.

2.6.11 Medicinal application of cerium

After its discovery, cerium found its use in gas mantles as well as in medical treatment. In 1854, James Simpson reported that cerium nitrate suppressed vomiting, particularly the one associated with morning sickness. It was also prescribed for the treatment of sea sickness and cough suppressant for patients of tuberculosis. In the last century, even over-the-counter medicines such as Novaurin and cerocol were on sale. It is reported that a weak solution of cerium nitrate is found to be an effective treatment for bathing the skin of people suffering from extensive third-degree burns, which is now a standard procedure in burn units.

Recent research on cerium oxide [23] reveals that cerium oxide nanoparticles or nanoceria possess the host of biological activity, based on its ability to inactivate ROS and to scavenge free radicals. Both ceria and cerium ions are having this property so that they behave like phosphatase. Future promise of cerium lies in its application in drug delivery, treatment of the diseases such as those associated with oxidative stress, redox therapy of oncological diseases, as adjuvant in antiviral treatment, immunomodulator and even as a modulator of signal transduction in neurology.

2.6.12 Medicinal application of chromium

Although deficiency of chromium in the human body is rare, yet it is needed in trace amount by our body. There are references available in the literature regarding involvement of chromium in the metabolism of fat, carbohydrates and proteins [24–31]. Chromium is known to enhance the action of insulin.

Chromium picolinate is used in the treatment of diabetes and depression, although a more number of studies are needed to fully understand the relationship between its use and management of diabetes. Chromium has been identified as an active ingredient in glucose tolerance factor [32]. There are though uses of chromium as picolinate in alternative medicine to treat or aid in controlling of blood sugar, to lower cholesterol and also as a weight loss supplement, it is sold only as a supplement and not as a medicine.

2.6.13 Medicinal applications of cadmium

Cadmium and its compounds are although toxic in certain forms and concentration, the British Pharmaceutical Codex from 1907 mentions that cadmium iodide was used as medication for the treatment of enlarged joints, scrofulous glands and chilblains. There are hardly any medicines using cadmium. There is no established use of this toxic metal as either supplement or medication.

2.6.14 Medicinal/medical application of cobalt

Cobalt is used as part of vitamin B12 for the treatment of pernicious anaemia, since it improves the formation of red blood cells. It helps in management of fatigue, digestive problems and neuromuscular ailments. It is a replacement of manganese and Zn in activation of several enzymes. They are called biochemical reaction activators. It is also part of the biotin-dependent Krebs cycle, which is the process our body uses for breaking sugars to energy.

2.6.14.1 Toxicity and side effects

Cobalt is reported to be toxic to the heart muscles and can cause toxic cardiomyopathy, if one has too much exposure. Too much of cobalt intake may cause goitre and can reduce thyroid activity.

Cobalt-60 is used as a medical source of radiation, primarily for cancer radiotherapy. It is also used as a radioactive tracer in biology. Further, materials with 28–68% cobalt with chromium, nickel and molybdenum are used in dentistry. Owing to their strong resistance to abrasion as well as tarnishing and compatibility with mouth tissue, they are more useful as compared to other precious metals. They are also used in bone surgery.

2.6.15 Medicinal applications of Cu

The recorded medical use of copper is found in *Smith Papyrus*, the oldest books known. The book is an Egyptian medical text written between 2600 and 2200 BC. It records the use of copper in sterilization of chest wounds and also to sterilize drinking water. The *Ebers Papyrus* written in 1500 BC also mentions the use of copper as medicine [33].

Copper compounds show vast biological action such as anti-inflammatory and anti-proliferative biological action such as biocidal as well as suitability for radiotherapy. Although it has not found many uses, it has potential for development as a drug. It is commonly used in contraceptive intrauterine devices [34], and the number of other medical applications.

2.6.16 Medicinal application of iron

A number of diseases such as cancer, kidney problems and HIV/AIDS can cause anaemia. Taking iron with other medications can help treat anaemia. Ferrous malate, ferrous gluconate and even ferrous sulphate are used to prevent anaemia. A number of iron supplements are available with brand names like Auryxia, Beef iron wine, Bifera, Elite Iron, Femiron, Feosol, Fergon, Hemocytes, Fer-in-sol. Variety of iron supplements are presented for problems related to intestinal diseases, stomach haemodialysis, bleeding and burns.

2.6.17 Medicinal applications of gallium

Gallium compounds have gained significance over last few decades in the field of medicine (stable gallium nitrate and radioactive gallium are compounds of choice for diagnostic and therapeutic agents). For the treatment of calcium and bone metabolism disorders in cancer patients, the drug is useful [36]. They have also shown potential as anti-inflammatory and immunosuppressive agents in animal models of human diseases. Radioactive gallium ^{67}Ga citrate has been used as a tumour imaging agent. This is reported that radioactive ^{67}Ga citrate injected into tumour-bearing animals concentrates in malignant cells [37]. There are a number of studies reported on anti-neoplastic activity of gallium nitrate in cancer treatment, malignancy-associated hypercalcaemia, bone metabolism and immunosuppressive and anti-inflammatory agents. However, there is scope of further research for drug potential of gallium compounds.

2.6.18 Medicinal application of mercury

Although mercury is known as a toxic metal, yet in spite of its negative image it has been playing a positive and contributory role in the area of medicine as well as dentistry. It has many therapeutic uses in dental fillings, cosmetics, paints, ointments apart from application in thermometals and other devices of medical applications [38]. However, we know that mercury or its compounds due to their toxicity are still far from being used as potential medicinal candidates.

2.6.19 Medicinal application of iridium

Iridium is mainly used as its radioactive isotope iridium 192 produced by non-radioactive iridium metal. The half-life of this isotope Ir-192 is 73.828 days. It decays in beta particles and gamma radiation. It is commonly used for the high-dose rate brachytherapy, procedure that involves placing radioactive material inside your body allowing a doctor to deliver higher dose of radiation to more specific body area. For the treatment of different types of cancer, Ir-192 implants have also been used. As they have curative properties, they are primarily used as implants in breast and head. The implants made of wires are introduced into the target area through a catheter. The implant wire is removed after the time required to deliver the desired dose. The procedure is very effective for providing localized radiation to the tumour site and avoids the whole body of patients from radiation. The isotope Ir-192 has also been used in treatment of prostate cancer.

2.6.20 Medicinal application of Mg

Magnesium is the fourth most abundant essential mineral in our body, half of it is distributed in bone and other half in the muscles. It is reported to be present less than 1% in our blood [39]. Common magnesium formulation available as supplements is magnesium oxide (MagOx) which contains about 61% elemental magnesium and sold as tablets. The other common formulation is magnesium hydroxide or milk of magnesia containing about 42% elemental magnesium available as suspension. It is prescribed for the treatment of a number of diseases such as eclampsia and pre-eclampsia. It is also recommended for the treatment of torsades de pointes and rapid atrial fibrillation through intravenous injection. Parenteral magnesium is also recommended for severe acute asthma, which improves peak expiratory flow rate and forced expiratory volume in 1 s [40]. It is also useful as oral and parenteral administration in improving symptoms of migraine. It is widely accepted and effective for the treatment of dyspepsia. Further, it is accepted as a standard treatment for constipation. An excellent review on medicinal application of magnesium is available in the literature, explaining pharmacology and mechanism of action of magnesium [41].

2.6.21 Medicinal application of molybdenum

Although deficiency of molybdenum is very uncommon, yet it is known to work in the body to break down proteins and other substances. It is most commonly used for managing molybdenum deficiency. There are recommendations for treatment of cancer of oesophagus and Wilson's disease by molybdenum. There are unsubstantiated claims regarding treatment of arthritis, cancers and neuropsychiatric disorders [42]. Literature survey reveals the use of molybdenum compounds in various applications (www.imoa.info/HSE/env). These applications can be listed in brief in Table 2.7.

Table 2.7: Some applications of molybdenum compounds.

S. no.	Compound	Application
1	Mo nanoclusters	Biological imaging
2	MoS_2 (polyaniline–MoS_2 hybrid nanostructure)	Cancer biomarker
3	MoS_2 (dual-responsive molybdenum/copper sulphide-based delivery system	Enhanced chemophotothermal therapy.
4	MoS_2 (tumour-targeted biocompatible nanodots albumin nanospheres)	Dual modality therapy agent for synergistic photothermal radiation
5	MoO_3 nanoparticles	Bactericidal
6	MoO_3 (polycaprolactone nanofibre composites)	Skin cancer
7	MoS_2 nanoparticles	Alzheimer's disease

2.6.22 Medicinal application of manganese

Manganese is used for the prevention of diseases such as manganese deficiency, osteoporosis and anaemia. It is also useful in the management of symptoms of premenstrual syndrome. Manganese though needed in our body in very small amounts, yet it is important in variety of functions like building bones, healing wounds and help in use of carbohydrates and amino acids in our body. It is mostly found concentrated in mitochondria of cells. It is available both as single supplement and part of many multivitamins. Manganese gluconate, manganese sulphate and manganese citrate are some single supplement prescriptions of manganese.

2.6.23 Medicinal application of osmium

No medicinal application of osmium or its compounds is referenced. However, in alternative medicines, osmium metal cum 30 is mentioned as medicine for glaucoma.

An alloy of platinum and osmium is used in surgical implants like pacemakers as well as heart valves, which shows its medical use.

2.6.24 Medicinal application of palladium

It is reported that palladium can be a future player in the field of medical sciences. Palladium has a significant role in medical/health care. Palladium-103 is employed as an alternative to prostatectomy using brachytherapy. The procedure involves putting seeds (of the size of rice) or titanium castings filled with palladium-103 implanting into the prostate. This is absolutely a non-invasive procedure. Apart from this, it has found use in blood glucose testing (part of electrode).

Palladium nanoparticles have of late been the focus of medicinal chemistry, and this application is growing interest. There is a need to address their unique properties and low toxicity advantage. There are research publications regarding their applications for targeted prodrug activation and cancer photothermal therapy as antimicrobial and cytotoxic agents to mention a few [43].

2.6.25 Medicinal/medical application of platinum

A review on medical application of platinum enumerates various useful and future applications of platinum [44]. Apart from the use of platinum as anti-cancer drugs such as cisplatin and carboplatin, variety of applications of platinum include its use for making components of medical appliances such as pacemakers, unplantable defibrillators, catheters, stents and neuromodulation and biomedical devices. Table 2.8 lists the application areas where platinum is used.

Table 2.8: Medical applications of platinum.

S. no.	Application	Device
1	Pacemakers, defibrillators, hearing devices, heart pumps	Electromedical implants
2	Stents, angioplasty ablation, distal protection	Interventional
3	Spinal fixation implants of hip and knee	Orthopaedics

Using platinum-based drugs, there has been remarkable advancement in the survival rate of cancer patients. It is estimated that 98% of testicular cancer patients will survive for 10 years after diagnosis using the anticancer drug cisplatin [45]. The other drug carboplatin is also used for other cancers such as ovary, breast and lung and is quite successful [46].

2.6.26 Medicinal uses of antimony

Some of the antimony-based compounds have beneficial effect such as meglumine antimoniate and sodium stibogluconate. The two compounds are principal medications used to treat leishmaniasis. This disease is caused by infection by a protozoan parasite. The disease is called Kala-azar. Although the drugs are still first line, they are extremely limited due to their side effects.

2.6.27 Medicinal application of vanadium

Medicinal application of vanadium is found in the treatment of a number of diseases including diabetes, low blood sugar, high cholesterol, heart disease, syphilis, tuberculosis and anaemia. It is also used as a supplement to improve athletic performance in weight training.

2.6.28 Medicinal/medical application of Zn

Zinc is important for growth and body tissues. It is also used to prevent zinc deficiency. Zinc is second to iron as the most abundant trace mineral in our body, and the commonly marketed zinc supplements are zinc gluconate, zinc acetate, zinc sulphate, zinc picolinate, zinc orotate and zinc citrate. A number of cold lozenges contains zinc, and a variety of other cold remedies also have zinc. It is involved in different aspects of cellular metabolism. It plays a vital role in the catalytic activity of about 100 enzymes. It also has a significant role in immune function, protein system, wound healing and many other related to development and growth.

The above description of individual application of metals in medicine shows how important metals are in medicinal chemistry. The basic objective of the chapter is to highlight the contribution of medicinal inorganic chemistry which is hitherto underplayed and less recognized. There is a need to not only recognize importance of this branch but consolidate the research area and focus this field for future welfare of medicinal chemistry. This would be an important branch for human welfare and fighting disease. A summary of various elements of periodic table having established or being explored as medicinally important is being given to have a state of the art of medicinal inorganic chemistry.

Chapter 3
Metals in alternative system of medicine

3.1 Ayurveda

Ayurveda is an alternative system of medicine, which has been time tested for about 5,000 years or more. It was found by ancient seers that drugs of different origin such as metal, animal or herbal are suitable for maintaining health in healthy individual and removing disease in a diseased person. The Sanskrit word 'Ayurveda' is originated by combining two words 'ayur' meaning life and 'veda' meaning knowledge. Its materia medica comprises resources in the form of drugs derived from plants, animals or even metals as well as mineral resources. Thus, application of metals or their metallic preparations is understandably a unique advantage of this alternative medicine system. However, scientific community has shown concern regarding safety of Ayurvedic herbal, herbo-mineral and metallic preparations which is a major issue for the age-old heritage [47]. However, this chapter of the book is an attempt to critically and cautiously present state-of-the-art application of metals or their preparations to medicinal chemistry only to strengthen the knowledge of medicinal inorganic chemistry and its significance.

The preceding section of this chapter mainly describes the use of metals/metallic preparations as medicine in Ayurveda, that is, the traditional Indian system of alternative medicines. To introduce the terminology, a glossary of words which includes appropriate or most close meaning is also being given to make the reader conversant with it.

3.2 Ayurveda (ayur + veda) life: knowledge

- *Rasashastra* – it is an integral part of Ayurveda dealing with the drug of mineral origin. This comprises their processing, characteristics, therapeutic use and toxicities along with their management.
- *Charaka Samhita* – one of the scheduled books of Ayurveda
- *Sushruta Samhita* – Ayurvedic classic written before *Rasashastra*

Historically, first records of useful drugs in India were maintained in Ayurveda, and Acharya Charak, the great Ayurvedic scholar (1500 BC), has mentioned various types of drugs based on the source. The pharmacology of drugs has been described with precision in these texts [48]. Metallic medicines have been an integral part of this alternative system of medicine, so the relevance of this chapter is more enhanced.

https://doi.org/10.1515/9781501516115-003

3.3 Superiority of mineral drugs

The alchemy and medicines based on metallic preparations were superior in comparison to treatment based on vegetable drugs or surgical therapies. There are mentioned in texts [49] based on classification of therapies into three categories, viz., *asuri* (demonic) includes surgical therapies, *manusi* (human) being performed using decoctions of vegetable drugs and lastly *daivi* (divine) which is performed using metallic and mineral preparations. It is further highlighted that mineral remedies are even in small doses therapeutically more effective unlike vegetable products generally administered in much larger doses. The mineral products are effective instantaneously as they do not have to pass through the process of digestion and metabolism before they are therapeutically active.

3.4 Concept of health in Ayurveda according to Sushruta [Sushruta, Sutra 15: 40]

A person having equipoise of *dosas* (factors that control physiological activities of the body), *agnis* (factors that are responsible for digestion and metabolism), *dhatus* (tissues, elements), *malas* (excreta) and, finally, *Kriya* (physical and mental activities) and the person who possesses spiritual, sensual and mental happiness are called *svastha or healthy*. Thus, the concept of health in Ayurveda is holistic rather than narrow.

Dhatu (metal) word of Sanskrit originates from a verb dha meaning holding, maintaining or supporting [50]. In Ayurveda, *Dhatu* means metals. Without going into details of the system, the chapter is based on description of references signifying the use of metals for medicinal and similar purposes. The *Charaka Samhita* is the basic text which gives such details. It mentions three major categories giving an insight about the use of metals:

(i) Utilization in therapeutics internal as well as external application
(ii) Application in equipment or instruments
(iii) Other purposes

3.5 Various metals used in the Ayurvedic system of medicine

3.5.1 Gold or aurum; au (Svarna)

Mention of gold as per *Charaka Samhita* is for analgesic, anti-oxidant, anti-depressant, anti-anxiety and cataleptic activities. Further, it is also useful as an immunomodulator. There are references available for analgesic activity of *Swarna Bhasma*, *Kushta Tela Kalan* and *Auranofin*, all Ayurvedic gold preparations regarding this application [51]. The Bhasma is the form of gold in the metallic state. The quantitative analysis reported for Bhasma is a combination of metallic gold (96.76%), silica (1.14%), ferric oxide (0.14%), phosphates (0.78%), potash (0.16%), salt (0.078%) and traces of copper and magnesium [52].

Gold has been used as an anti-pruretic agent to relieve itching of palms. Robert Kochs observed in vitro anti-bacterial activity of gold against. *Mycobacterium tuberculosis*. The gold compounds affect factors that influence the immunological response [54] and decrease the concentration of rheumatic factors also [53].

Gold suppresses the anaphylactic release of histamine more effectively than glucocorticoids [54]. A water-soluble preparation using sodium aurothiomalate has been established to treat arthritis, which is administered through IM injections. Although its pharmacokinetics has not been established and as reported its effects are due to antimicrobial activity and stimulation of reticuloendothelial system [55], conclusively it can be mentioned that Ayurvedic system has significant use of gold in its medicines.

3.5.2 Silver (Rajat)

Use of silver in Ayurvedic therapeutics also dates back to the time of Charaka. Ancient classics describes that although its use as medicine was less extensive compared to other metals like copper or iron (Tamra or Loha), yet it had an important place in the system of medicine. Silver used as Bhasma is a cooling astringent and has sour taste and is a laxative. It promotes vitality, strength and appetite [49]. Quantitative analysis of 'Rajat Bhasma' as reported [52] comprises metallic silver (52–59%), free sulphur (0.675%), ferric oxide (14.33%), calcium (10.769%) and silver chloride (0.479%) with traces of sodium, potassium and aluminium.

3.5.3 Copper (Tamra)

Copper is one of the oldest metals known to human civilization. Its application for day-to-day livelihood is an established fact. Not only this, it was used in alloys called brass and bronze. Charaka uses Arka as its synonym, which has been clarified by Chakrapani [56]. References are available on physico-chemical parameters of Tamra Bhasma, which can be briefly described as acid-insoluble ash 1.12%, pH 6.76, loss on drying 1.14%, Hg (10.9%), sulphur (1.50%), copper (61.48%) and bulk density 6.288 g/cc [57].

As described in Ayurvedic texts, it is useful in abdominal diseases including ascites, anaemia, piles, obstinate skin disease like leprosy, bronchitis, asthma, tuberculosis, chronic rhinitis and liver disorders. Its alloy brass is also reportedly useful as an Ayurvedic preparation for the treatment of parasitic infestations, leprosy and serious anaemias. Bell metal or *Kamsya* is another alloy of copper used in Ayurveda for the treatment of parasitic infestation and skin diseases [49]. In addition to these, miscellaneous copper compounds including copper pyrite (*Svarna Maksika*), iron pyrite (*Raupya Maksika*) and copper sulphate (*Tuttha*) have also been reported as medicinal compounds [49].

3.5.4 Iron (Aayasa or Loha)

After Svarna, Rajat and Tamra (Au, Ag and Cu), iron is a metal known to have medicinal use in Ayurvedic system. In various forms, iron compounds have been reportedly useful for treatment of (internal and external administration) pathological manifestations. Iron deficiency in anaemia can be treated using iron compounds explains this system. *Rasashastra* classics mention that iron is a rejuvenator and it stimulates the function of activities of all organs. It is reported to have haematinic and cytoprotective properties as *Lauha Bhasma* and *Mandura Bhasma* (iron-ash). It is reported to cure jaundice, colic, asthma, malabsorption, haemorrhoids, tuberculosis, polyuria, skin diseases and liver diseases [58].

3.5.5 Tin (Sn) (*Vanga*)

Tin named *Vanga* in Ayurveda has been known to Indian physicians since ancient times. It has been categorized in *Charaka Samhita* in two varieties *khuraka* and *mishrika*, out of which former has been accepted for therapeutic use [59]. The quantitative analysis as reported for *Vanga Bhasma* [7] is a combination of stannic oxide (91.4%), ferric oxide (2.9%), potassium (2.9%), calcium oxide (2%), aluminium (2%) and magnesium (0.6%) oxides.

Formations of Vanga are reported to be beneficial for different diseases like genitourinary tract. In Bhasma form, it is reported to be useful in the treatment of obstructed urinary disorders including diabetes and premature ejaculation.

3.5.6 Lead (Nāga)

Lead has been mentioned for its therapeutic application in Ayurveda in two forms: Naga Kumara and Naga Samala. Of the two kumara variety is acceptable for therapeutic use. Quantitatively, Naga Bhasma is a combination of lead oxide (75.6%), ferric oxide (7.5%) with traces of calcium and magnesium chlorides and carbonates. It has been reported to be useful for the treatment of menorrhagia, piles and urinary disorders and diabetes. Mention of other forms or compounds including red lead, litharge, lead sulphate also makes it amply clear that this metal has been used in Ayurvedic system of medicine.

3.5.7 Zinc (Yasada)

In Ayurvedic system, zinc or Yasada has been used as Yasada Bhasma to treat variety of diseases, particularly as anti-diabetic [60]. It has also been reported to have anti-oxidant [61] properties.

Miscellaneous metals or metallic preparation with therapeutic potential in Ayurvedic system attract attention. However, this book is attempting to simply highlight these metals/metalloids or preparation to provide possible exposure of this branch in a more holistic view and does not advocate or criticize any system.

3.6 Metals in Tibetan Zuotai

An excellent review on chemical composition of metals in Bhasmas as well as *Tibetan Zuotai* critically reports therapeutic effects vis-a-vis toxicity of Tibetan system of medicine [62]. Bhasmas are processed Ayurvedic medicines which are named *Zuotai* in Tibetan medicines. It is known that minerals are processed alchemically as Bhasmas in Ayurvedic medicines and Zuotai in Tibetan system of medicines. Herbo-metallic medicines in India have long been used for their medicinal properties. In Ayurveda metallic preparations mineral use contributes 8%, which includes Au, Ag, Cu, Fe, As, Al, lime, lead and zinc, not to mention compounds of these metals.

Similar to Bhasmas in Ayurveda, Zuotai is a mixture of metal ash included in Tibetan medicines [63]. The procedure to prepare Zuotai is similar to Ayurvedic Bhasmas. This takes about 3–4 months of repeated treatment and incineration procedure based on animal and herbal products. As reported in the review, Zuotai constitutes two ashes, viz. Nengchi Bakuang and Nengchi Bajin [64, 65]. The Nengchi Bakuang contains chemical components from mica like SiO_2, $CaCO_3$, $K_2Ca(SO_4)_2$, H_2O, KCl and also some metals like $FeAS_2$, FeAs, Fe_2As, Cu_2S and AsFe, while the other ash Nengchi Bajin comprises $AuPb_2$, PbO, $PbSO_4$, Ag_2S, CuS, SiO_2, CuO, FeS, SnS, though the major component of metal such as Zuotai is metacinnabar (54%) (β. Hgs.) [66].

3.7 Metals in traditional Chinese medicines

An excellent review on metals in traditional Chinese medicinal (TCM) materials [67] forwards a very exhaustive analysis of application of metals in the system.

TCM system has become an important support to Chinese health system in recent years. It is in a sense holistic natural medicinal system parallel to Ayurvedic system of India. The TCM system is one of the most systematic traditional medicinal systems. About 87% of raw materials in Chinese medicinal materials are derived from plant parts. To name roots, rhizomes, tubers, stems, flowers, fruits or even whole plants constitute raw material or TCM material [68], and according to a survey by the WHO about 60–80% of the world's population uses traditional or alternative medicines mainly from plants [69]. Although TCM material does not use metals directly as in Ayurvedic or Zuotai, indirect metals in them have been found, and its toxicity or advantage has been described in the literature. Thus, the present context of application of metals in this system does not warrant much discussion. However, from the point of medicinal application, the part has been included in the book.

3.8 How Bhasmas and Zuotai benefit

The beneficial effects of Bhasmas and Zuotai are explained on the basis of a phenomenon called Hormesis. Hormesis is a term in toxicology referring to a biphasic dose–response to an environmental agent characterized by a low-dose stimulation or beneficial effect with a high-dose inhibitory effect. Merriam Webster defines 'Hormesis' as a theoretical phenomenon of dose–response relationship in which something that creates a harmful biological effects at moderate-to-high doses can produce beneficial effects at low doses. That derives the corollary that even toxic heavy metals can be beneficial to humans at very low concentration showing activities along with its toxicity [70]. The Indo-Tibetan system of traditional medicines claim that proficiency in the suggested longevity practices includes meditation, diet and yoga. This enhances health effects. Even homeopathic medicines include metals. It is important to know that Bhasmas and Zuotai are used in addition to other herbo-metallic mixtures in quite small amounts. Understandably, low doses of processed minerals partly fit into Hormesis theory.

A more generalized mechanism for traditional medicines is termed adaptation. It is explained by imagining herbo-metallic mixture as a whole, and Herbogenomic as a tool to identify the key molecular pathways after the administration of herbo-metallic mixtures at various dose levels. It is said that these herbo-metallic preparation could enhance immune function, induce anti-oxidant pathways or even affect drug processing genes in the same way as 'programme the liver' [72] for producing beneficial effects. A summary of adaptation mechanism can be given to briefly understand the mentioned approaches:

(i) Many alternative medicines such as Ayurvedic [73] and Tibetan medicines [74] probably activate NrF2-antioxidant pathways reducing oxidative stress as activation of Nrf2 is universal protective means to fight toxic stimuli [75]. Nrf2 is a basic leucine zipper (b zip) protein that may regulate the expression of antioxidant proteins that protects against oxidative damage caused due to injury and inflammation.
(ii) Immunomodulatory and anti-inflammatory properties: It is reported that Ayurvedic Bhasmas [76] and Zuotai-containing Tibetan medicines [77] modulate the immune functions in order to give beneficial effects.

(iii) Modulation of metabolism:
It is reported that for common metabolic disorders like diabetes and hyperlipidaemia, Zn-based *jasad Bhasma* [78], gold-containing preparations *Shadguna Balijarita Makar dhwaja* [79] and lead-based *Naga Bhasma* [80, 81] are effective, although effects of herbo-metallic recipes on metabolic processes are not established yet.

(iv) Modulation of neurological function:
It is reported that Zuotai is effective in improving symptoms which are like depression in chronic mild stress in mice. It increases 5-HT levels of the brain [82]. Gold preparations restore stress-induced elevation in levels of brain catecholamines and 5-HT and impart beneficial effects [83].

(v) Tissue repair and regeneration:
Gold nanoparticles have been reported [84] to enhance self-renewal and pluripotency for tissue regeneration. However, how Bhasmas and Zuotai containing traditional medicines promote tissue repair is further a matter of research before it is established.

Traditional medicines follow a general philosophy that poisons can be used to attack poison, It is expressed by results that under pathological conditions the efficacy of herbo-metallic preparations overcomes their toxicity [85].

The description in former sections amply support the fact that in alternative system of medicines, particularly Ayurveda, Tibetan medicines and Chinese system there has been use and application of metal-, herbo-metal- and mineral-based drugs. However, the field of research needs lot of focus and investigation, although there is lot of research going on; yet it is an interesting branch which needs attention under the banner of 'Holistic Medicinal Inorganic Chemistry'.

References

[1] Thompson, K. H., Medicinal inorganic chemistry: An introduction, https://dioorg/, 2010, 10.1002047 0862106.la470.

[2] Therapeutic potentials of metals in ancient India: A review through Charaka Samhita, Galib, Mayur Barve, Mayur Mashru, Chandrashekhar Jagtap, B.J.Patgirl and P.K. Prajapati. J. Ayurveda Integr. Med., 2011, 2(2), 55–63.

[3] Chelating agents and metal chelates, Dwyer, F. P., Mellor, D. P., Academic Press, 1964.

[4] Meulenbeld, G. "10-Caraka, his identity and date" A History of Indian Medical Literature, Vik, 1A, Part 1, Forster, G. E., (1999), p. 114.

[5] Rossolti, J. C., H., Rossolti, The determination of stability constants, McGraw Hill, New York, 1961.

[6] Griffith, J. S., The theory of transition metal-ions, Cambridge University Press, London and New York, 1961.

[7] Orgel, L. E., an introduction to transition metal chemistry, Methuen, London, 1960.

[8] Jorgensen, C. K., Absorption spectra and chemical bonding in complexes, Pergamon, New York, 1962.

[9] Bhattacharya, A. A., S., Curry, N. P., Franks, Binding of general anaesthetis propofol and halothane to human sserum albumin, J. Biol. Chem., 2000, 275.

[10] Vincent, J. L., Revelance of albumin in modern critical care medicines, Bailliere's best practice and research, Clinical Anaesthesiology, 2009, 23(2), 183.

[11] Brillo, D. J., D. E., Marx, Silver in medicine a brief history BC 335 to present, Burns, 2014, 40, Suppl 1, 53–58, 10.1016/J.burns 2014.09.009.

[12] Alexander, J. W., History of medical use of silver, Surg. Infect. (Larchmt), 2009, 10(3), 289–292, 10.1089/Sur.2008.9941.

[13] Alan, B. G., Lansdown, Advances in pharmacological and pharmaceutical sciences, 2010, http:/ doi.org/10.1155/2010/910686.

[14] Silver in health care: Its antimicrobial efficacy and safety in the, Alan, B.G. Lansdown, http://org/10.1039/9781849731799, 2010.

[15] Higby, G. J. Gold in medicine. Gold Bull, 15, 130–140 (1982) http://doi.org/10.1007/BF03214618.

[16] Voegtlin, C., A., Dyer, C. S., Leonard, U.S. public health repts, Vol. 38, 1882, 1923.

[17] Antman, K. H. Introduction: The history of arsenic trioxide in cancer therapy. The oncologist 2001; 6 (Supplement 02): 1-2 (Pub Med) Google Scholar. Stephen John Ralph, Metal based drugs 2008: 260146 doi 10.1155/2008/260146.

[18] Stephan, J. R., Metal based drugs 2008: 260146 doi. 10.1155/2008/260146.

[19] Antman, K. H., Introduction: The history of arsenic trioxide in cancer theapy. The Oncologest 2001, 6 (Supplement 02): [PubMed] Google Scholar.

[20] Salvador, J. A., S. A., Figueiredo, R. M., Pinto, S. M., Selvestre, Bismuth compounds in medicine chemistry future, Med. Chem., 2012, 4(11), 1495–1523, 10.4155/fmc.12.95.

[21] Keogan, D. M., D. M., Griffith, Current and potential application of Bismuth based drugs, Molecules, 2014, 19(9), 15258–15297, 10.3390/molecules, 190915258.

[22] Cheng, H. U. L., T., HeB, L., Li, Y., Wang, Y.-T., Lai, G., Jiang, H., Sun, Identification of medal associated proteins inc ells by using continuous flow gel electrophoresis and induchrely coupled plasma mass spectrometry, Angew. Chem. Int. Ed, 2013, 52, 4116- 49-20, 10.1002/anie, 201300794 [Pub Med.

[23] Shcherbako, A. B., Y., Nadezhda, M., Zholobak, V. K., Ivano, Biological, biomedical and pharmaceutical application of cerium oxide https://doi.org/10.1016/B 978-0-12-815661-

https://doi.org/10.1515/9781501516115-004

2.00008-6. (Chapter of cerium oxide (CCO_2). Synthesis, Properties and Applied Metal Oxide (Science, Direct) Elsevier.

[24] Mertz, W., Chromium occurrence and function in biological systems (1969), Physiol. Rev., 49, 163–239.

[25] Mertz, W. Chromium in human nutrition: A review (1993): 626–633.

[26] Jr, H., L. L., Ransome, Kufio, Majaj A.S., Improvement of impaired carbohydrate metabolism by chromium (III), Am. J. Clin. Nutr., 1968, 21, 203–211.

[27] Jeejeebhoy, K. N., R. C., Chu, E. B., Marliss, G. R., Greenberg, A., Bruce Robsertson, Chromium deficiency glucose intolerance and neutropathy reversed by chromium supplementation in a patient receiving long-term total paranteral nutrition, Am. J. Clin Nutr., 1977, 30, 531–538.

[28] Chromium, A. R., Trace elements in human and animal nutrition, edited by, Mertz, M., Academic Press, San Diego, CA, 1987, 225–244

[29] Lukaski, H. C., Chromium as Supplement, Annu. Rev. Nutr, 1999, 19, 279–302.

[30] Stoecker, B. J., Chromium, Present knowledge in Nutrition, 8th, Edited by, Bowman, B., Russel, R., IL 51 Press, Washington, DC, 2001, 366–372.

[31] Vincent, J. B., The potential value of toxicity of chromium picolinate as a nutritional supplement, weight loss agent and muscle development agent sports, Med, 2003, 33, 213–230.

[32] Schwarz, K., W., Mertz, Chromium (III) and the glucose tolerance factor, Arch. Biochem. Biophys., 1959, 85, 292–295.

[33] Dollwet, H. H. A., J. R. J., Sorenson, Historic uses of copper compounds in medicine, Trace Elem. Med., 1985, 2(2), 80–87.

[34] Urtiz, M. E., H. B., Cro Xatto: Copper T-intra uterne device and elvonorgestrel int-rauterine system: Biological bases of their mechanism of action contraception, 2007, 75 (6 Suppl) 5- 16-30(Pub Med: 17531610].

[35] www.webmd.com/vitamins and supplements.

[36] Chitambar, C. R., Medical Applications and Toxicities of Gallium Compounds; Int, J. Environ Res. Public Health., 2010, May 7(5), 2337–2361.

[37] Edvards, C. L., R. L., Hayes, Tumor scanning with 67 Ga/Citrage, J. Nucl. Med., 1969, 10, 103–105, Pub Med] [Google Scholar.

[38] Iqbal, K., M., Asamt, Uses and effects of mercury in medicine and dentistry, J. Ayub Med. Coll Abbottabad, 2012, 24(3–4), 204–207.

[39] National Institute of Health Office of Dietary Supplements. Magnesium: http://ods.od.nih.gov/factsheets/magnesium.asp.

[40] http://www.aafp.org/afpsort.xml.

[41] P., Guerrera, M., S. L., Volpe, J. J., Mao, Am. Fam. Physician, 2009, July 15, 80(2), 157–162.

[42] https://www.urme.rochester.edu.co. (Google).

[43] Dumas, A., patrick couvreur palladium: A future key player in the nanomedical field, Chenise, 2015, Apr. 1, 6(4), 2153–2167.

[44] Cowley, A., A Healthy Future: Platinum in Medical Applications, Platinum Metals Rev., 2011, 55(2), 98, 10: 1595/147106711X 566816.

[45] Wihshaw, E., Platinum Metals Rev., 1979, 23(3), 90,Link http://www.technology.matthey.com/article/23/3/90–98.

[46] Barnard, C. F. J., Platinum. Metals Rev., 1989, 33 (4), 162 Link http://www.technology.Mathey.Com/article/33/4/162–167.

[47] Saper, R. B., S. N., Kales, J., Paquin, M. J., Burns, D. M., Eisenberg, R. B., Davis, et al., Heavy metal contents of Ayurvedic herbal medicine products, JAMA, 2004, 292, 2868–2873.

[48] Agnivesha, Quest for longevity, Samhita, C., Sharma, R. K., Dash, B., Chowkhamba Sanskrit Series Office, Varanasi, 2001.

[49] Vaidya Bharwan Dash. Alchemy and Metallic Medicines in Ayurveda, Introduction pp. 18-19, Concept Publishing Company (P.) Ltd, 1986.

[50] Monier-Williams Sanskrit - English Dictionary. http://sanskritdictionary.com.

[51] Bajaj, S., S. B., Vohra, Gold preparations in Ayurveda and Unani-Tibb, therapeutic rationale and safety aspect, J. Trace Elem. Exp. Med., 1998b, 59, 187–193.

[52] Jha, C. B. 'Ayurvedic Rasashastra' Varanasi India: Choukhamba Surabharati Prakashan, 2000. p. 314.

[53] Barve, G. M., M., Mashru, C. S., Jagtap, B. J., Patigiri, P. K., Prajapati, Therapeutic potentials of metals in ancient India: A review through Charaka Samhita, J. Ayurveda Integr. Med., 2011, 2(2), 55–63.

[54] Norn, S., Anaphylactic histamine release and influence of antirheumatics, Acta Pharmacol. Toxicol. (copenti), 1971, 30, 1–59.

[55] Goodman and Gilman, The Pharmacological Basis of Therapeutics, Sec. 5, Chap. 29, Macmilan Publishing Co, New York, Vol. 714, 1980.

[56] Caraka 'Caraka Samhita', Chik 141sa Sthaana 7/86, Varanasi India: Choukhamba Sanskrit Sansthan: 2000.

[57] Dahikar, G. K., S. B., Kamble, Shashi Prabha Gupta pharmaceutical and analytical study of Tamra Bhasma: A research article, Internat. Ayrvedic Med. J., 2019, 7, (12).

[58] Garg, N., A., Singh, A Text book of Rasa-Shastra, Chankabha Orientalia, Delhi, 2018.

[59] Rao, G. P. A Text Book of Bhaisajya Kalpana Vijnanam. The Mohandas Indological Series 13.

[60] Santhosh, B., P. G., Jadar, Samanya, Sodhna of parad (mercury) and its Namburi Phased Spot Test, Aryavaidyan, 2013, 26(2), 94–96.

[61] Liu, J., F., Zhang, R. K., Velaga Pudi, A., Olumayokun, C. L., Olajide, L.-X., Wei, Chemical Composition of Metals in Bhasmas and Tibetan Zuoati are a Major Determinants of their Therapeutic Effects and Toxicity, Evidence based complimentary and alternative medicine, Vol. 2019, https:/doi.org/10.1155/2019/1697804.

[62] Huang, H. B., Q. Z., Wang, X.-W., Wang, et al, Overview of current researches on Tibetan medicines "Zuotai", Zhongguo Zhongyao Zazhi, 2013, 38(17), 2886–2888.

[63] Li, C., L., Suo, G., Ze Wang, Y. Z., Du, L. X., Wei, Chemical and structural analysis of Nengchi Bakuang ashes in refining of Tibetan medicine Zuotai, Chin. J. Pharm.l Anal., 2011, 31, 2220–2224.

[64] Li, C., J., Duo, D., Zhan, et al, Chemical and Structural analysis of Nengchi Bajin ashes in refincing of Tibetan medicine of TSO thal, Zhongguo Zheng Yao Za Zhi, 2012, 37(13), 1952–1957.

[65] Li, C., H., Yang, Y., Du, et al., Chemical species, Micromorphology and XRD finger print. Analysis of Tibetan Medicine Zuotai containing mercury, Bio. Inorg. Chem. Appli., 2016, Article ID 7010519.

[66] Chen, Y., Z., Junyi, H., Sun, J., Qin, J., Yang, Metals in Traditional Chinese medicinal materials (TCMM): A systematic review, Ecotoxicology and Environmental Safety, 207, 1–36, 2020 (http://doi.org/10.1016/uj.ecoenv.2020, 111–311.

[67] Chu, Z., H., Chen, Y., Zhai, Chinese medicine resources, Liaoning Science and Technology Press, 2010, Google Scholar.

[68] Li, J., Y., Wang, H., Yang, P., Yu, Y., Tang, Five heavy metals accumulation and health risk in a traditional Chinese medicinal cortex moutan collected from different sites in China, Hum.

Ecol. Risk Assess, 2018, 24, 2288–2298, 10.1080/10807039.2018.1459181scopusgooglescholar.
[69] Nascarvella, M. A., E. J., Calabrese, A method to evaluate hormesis in nanoparticle dose response, Dose Response, 2012, 10(3), 344–354.
[70] Kang, Y. J., Herbogenomics from traditional Chinese medicines to novel therapeutics, Exp. Biol. Med., 2008, 233(9), 1059–1065.
[71] Klassen, C. D., Learning to program the liver, Annu. Rev. Pharmacol. Toxicol., 2014, 54(1), 1–8.
[72] Kishor, B., P., Rai, R., Bhatia, S., Kumar, S. K., Gupta, A., Singha, Adaptogenic potential of oxitard in experimental chronic stress and chronic unpredictable stress induced dysfunctional homeostasis in rodents, J Ayurveda Integr. Med., 2017, 8(3), 169–176.
[73] Wu, Q., W.-K., Li, Z.-P., Zhou et al. The Tibetan medicine Zuotai differs from $HgCl_2$ and MeHg in producing liver injury in mice, Regulat. Toxicol. Pharmacol., 2016, 78, 1–7.
[74] Velagapudi, R., A., El, Bakoush and O.A. Olajide, Activation of Nrf2 pathway contributes to neuroprotection by the dietary flavanoid tiliroside, Mol. Neurobiol., 2018, 55(10), 8103–8123.
[75] Chavare, A., P., Chowdari, S., Ghosh, et al, Safety and bioactivity studies of Jasad Bhasma and its in-process intermediates in Swiss Mice, J. Ethnopharmacol., 2017, 197, 73–86.
[76] Li, Q., H., Li, T., Xu, et al, Natural medicines used in the traditional Tibetan medical system for the treatment of liver diseases, Front Pharmacol., 2018, 9.
[77] Umrani, R. D., D. S., Agarwal, K. M., Paknikar, Anti-diabetic activity and safety assessment of Ayurvedic medicine, Jasada Bhasma (Zin ash) in rats, Indian J. Exp. Biol (IJEB), 2013, 51(10), 811–822.
[78] Khedekar, S., G., Rukkudin, B., Ravi Shanker, P., Prajapati, Anti-diabetic activity of traditional Indian gold containing preparations, Shadguna Balijirita Makardhwaja on strepotzotocin induced diabetic rats, J. Intercult. Ethnopharmacol., 2016, 5(2), 162–167.
[79] Nagarajan, S., K., Shivaji, S., Krishnnaswami, et al, Safety and toxicity issues associated with Lead-based traditional herbo-metallic preparations, J. Ethnopharmacol., 2014, 51(1), 1–11.
[80] Rajput, D., B., Patgiri, R., Galib, P., Prajapati, Anti diabetic formulations of Naga Bhasma (Lead CalX), A brief review, Ancient Science of Life, 2013, 33(1), 52.
[81] Zhao, J., C., Niu, J., Wang, et al, The depressive like behaviour of chronic unpredictable mild stress treated mice, ameliorated by Tibetan medicine, Zuotai: involvement in the hypothalamic pituitary - adrenal (HPA) axis pathway, Neuropsychiatr Dis. Treat., 2018, 14, 125–141.
[82] Shah, Z. A., R. A., Gilani, P., Sharma, S. B., Vohora, Attention of stress - elicited brain catecholamines serotonin and plasma corticosterone levels by calcined gold preparations used in Indian system of medicines, Basic Clin. Pharmacol. Toxicol., 2005, 96(6), 469–474.
[83] Paul, W., C. P., Sharma, K. D., Deb, Nature vs Nature: Gold perpetuates stemmess, J. Stem Cells, 2011, 6(4), 181–198.
[84] Liu, J., L.-X., Wei, Q., Wang, et al, A review of Cinnabar (HgS) and /or realgar (As4S4) containing traditional medicines, J. Ethnopharmacol., 2018, 210, 340–350.

Section B **Biological inorganic chemistry**

Chapter 1
Introduction

Table 1.1 shows the relative distribution of the most important chemical elements in lithosphere, atmosphere, hydrosphere and biosphere.

Table 1.1: Relative distribution of the most important chemical elements (%).

Element	Lithosphere	Atmosphere	Hydrosphere	Biosphere
O	61.1	21.1	33.2	24.9
Si	20.4			
Al	6.3			
H	2.9		66.2	49.7
Ca	2.1			
Na	2.1		0.3	
N		78.4		0.3
Ar		0.5		
Cl			0.3	
C				24.9

A living organism contains a wide variation of bioactive chemical elements, and their compounds assist the organisms' functions and help animals and plants live healthy [1–7]. The most important chemicals are mostly compounds that consist of combination of C, H and O atoms (organic compounds) with minor amounts of other chemical elements (Table 1.2). Of the known more than 100 elements, about 28% occur in the biological flora and fauna. It has to be mentioned that approximately most of the chemical elements of all groups are represented in biological materials though some of them are in trace amounts. The vital C in cooperation with N, H and O (the so-called major elements) are the most important for natural life, making up 96.3% of the weight of the body. The lasting 3.7% contain only seven additional chemical elements (Table 1.2), which are in trace amounts in many living organisms and can also be considered as essential for life. The group of minor elements consist of calcium, phosphorus, potassium, sulphur, sodium, chlorine and magnesium. In addition to these major and minor chemical elements, there are numerous trace elements which are mandatory in very small quantities (not >0.01%) usually participating in different components of biomolecules like

https://doi.org/10.1515/9781501516115-005

enzymes and are needed for their activity. The group of trace elements contain aluminium, arsenic, boron, bromine, chromium, cobalt, copper, fluorine, iron, manganese, molybdenum, selenium, silicon, tin, vanadium and zinc.

Table 1.2: Abundance of elements in the human body.

	Element (atomic symbol)	Abundance (% wet weight)
Major chemical elements	C, N, H, O Total	96.3
Minor chemical elements	Ca, P, K, S, Na, Cl, Mg Total	3.7
Trace chemical elements	Al, As, B, Br, Cr, Co, Cu, F, Fe, Mn, Mo, Se, Si, Sn, V, Zn Total	0.01

Chapter 2
Main-group elements

Group IA

The elements located in group IA of the periodic table are called alkali metals (lithium, sodium, potassium, rubidium, caesium and francium). Their physical properties are common to those of other metals, except the densities. Their single electron in the outer shell is loosely bound which determines their largest atomic radii, the lowest ionization energy, the lowest electronegativity and the highest reactivity. The alkali metals form compounds in which they have a constant oxidation state (+1) as univalent cations.

2.1 Hydrogen

Hydrogen belongs to group IA. Hydrogen and oxygen are the most essential living elements. In combination they form water, which is about three-fifth of the human body, thus making life possible [4]. Water liquifies the life-supportive compounds and transfers them to cell fluids. Most of the vital bioreactions take place in water. Hydrogen is visibly a crucial component of water. It forms hydrogen bonds, giving water most of its exceptional features. Likewise, hydrogen binds to carbon forming the construction of bodies. Stomach acid for digesting the food and absorbing the other necessary elements for the survival is also composed of hydrogen and chlorine. Numerous biochemical reactions which make life possible include H^+ cation.

2.2 Lithium

Lithium has extensive usage as a healing agent in the treatment of depressive disorders [6, 7]. This element also causes various symptoms connected with the nervous system like psychosomatic retardation, as well as neuromuscular and cardiovascular symptoms. In the body, lithium exists as Li^+ cation. Its poisonous effects are probably as a result of its resemblance to biologically essential sodium and potassium ions. Lithium cation is a nervous system poison that causes also kidney damage [8].

Its prevalent occurrence in plants results in a widespread distribution in animal organisms and human beings. Lithium inorganic salts are not extremely poisonous. Numerous lithium activities are critical for the determination of its healing effectiveness [9].

Lithium is predisposed to form carbon–metal bonds in organometallic compounds with covalent properties. These organolithium compounds are moisture penetrating and could undertake ignition in air conditions. Some of the most important

https://doi.org/10.1515/9781501516115-006

organolithium compounds are ionic, such as methyllithium ($LiCH_3$), ethyllithium (LiC_2H_5), *tert*-butyllithium (LiC_4H_9) and phenyllithium (LiC_6H_5) [5–7]. Lithium also produces a carbonyl (LiCOCOLi) with high toxic effects.

2.3 Sodium

Sodium exists in almost all foods. Sodium inorganic salts, predominantly NaCl, occur everywhere in the biological environment. Sodium and potassium are essential bioelements that maintain a fixed balance in the cells. The biological effects are mainly connected with the sodium cation. Sodium cations can substitute Ca^{2+} in biomolecules and complexes. The smallest Na requirement is about 0.05% of the nutrition, equivalent to a necessity of 1–2 g of salt per day. Sodium content in tissues varies, for instance, around 0.62% in blood and 0.1% in skin. Substantial amounts of sodium can be lost by the skin sweating, and significant quantities are excreted by the urine. The most important sodium salts for humans are chloride and carbonate [4, 5]. Without Na, the cells cannot get the needed nutrients. This element permits the body to keep the optimal blood chemistry and allows the normal contracts of muscles and the regular nervous system functioning.

2.4 Potassium

Potassium is a very significant element to cells, and we could not survive without it. Each organism has a closely preserved level of potassium and a moderately fixed K–Na ratio. It is the main cation within the cells, whereas Na cation is the main constituent of the extracellular liquids. Potassium, nitrogen and phosphorus are the most essential elements in the plant realm, which are known as key plant nutrients. The content of potassium in plants differs significantly, normally between 0.5% and 2%. The potassium proportion between the content within the cells and that of extracellular plasma is around 27:1. The K content in muscle tissues is about 0.3%, while in the blood serum it is approximately 0.01–0.02%. The nutritional necessity is roughly 3.3 g per day. Superfluous potassium can be excreted by the urine and by sweating [4, 5].

Sodium and potassium cations cross the cell membrane with the purpose of adjusting the intracellular and extracellular concentrations. This transmembrane transportation is passive or active by the usage of the so-called Na^+/K^+ pump. The transport occurs along ion membrane channels. The required energy for the transportation

$$Na^+\,(in) \rightarrow Na^+\,(ex)$$

$$K^+\,(ex) \rightarrow K^+\,(in)$$

is given by the ATP hydrolysis [5, 7].

Organometallic compounds of sodium and potassium have less covalent carbon–metal bonds than the equivalent lithium bonds and are more reactive. Organometallic compounds of caesium and rubidium are rare, so their toxicologic implication is comparatively negligible [5, 7]. Consequently, only the toxicologic effects of Na^+ and K^+ compounds are of interest. Na^+ and K^+ ions and their compounds are not poisonous at standard biological levels. Their oxides and hydroxides are caustic and corrosive and may damage the contact tissues. Oxides can be obtained by burning their organometallic compounds, and the respective hydroxides can be produced by direct interaction of organometallic compounds with water:

$$C_5H_5 - Na^+ + H_2O \rightarrow C_5H_6 + NaOH$$

Sodium and potassium can produce also carbonyl compounds (with the formulas NaCO and KCO), which are very reactive. By the decomposition of the carbonyl compounds along with the respective oxides and hydroxides, the toxic CO can also be obtained.

Alkoxide compounds of sodium and potassium M^+–OR can be obtained by the reaction

$$2CH_3OH + 2Na \rightarrow 2Na^+ \; {-OCH_3} + H_2$$

The methoxide is extremely caustic, and in the interaction with water forms the hydroxide:

$$K^+ \; {-OCH_3} + H_2O \rightarrow KOH + CH_3OH$$

Chapter 3
Main-group elements

Group IIA

The chemical elements located in group IIA are called alkaline earth metals. Like group IA elements, alkaline earth metals have typical metallic properties – low electron affinity and electronegativity. They have two valence electrons and consequently smaller atomic and ionic radii. In their compounds, the alkaline earth metals are in the form of bivalent cations.

3.1 Beryllium

The first member of the group 'beryllium' is remarkably poisonous. That is why the contemporary usages of Be necessitate careful handling practices to avoid the contact. This element has numerous poisonous effects, especially on the skin causing dermatitis and conjunctivitis. Beryllium inhalation causes pneumonitis with inflammatory reactions of the whole respiratory tract. Beryllium fluoride (BeF_2) is mostly toxic [5–7].

The beryllium chemistry is different from that of alkali and alkaline earth metals because of its smallest atomic and ionic radii. All this reproduces a high polarizing ability; consequently, beryllium tends to form covalent but not ionic compounds, as well as coordination complexes. The capability of Be to produce chelates can be used in beryllium intoxication treatment [10].

3.2 Magnesium

Magnesium is essential for many different biological functions [4, 5], including growth and formation of the muscles and bones. Mg stops heart syndromes and regulates the blood pressure. This element is used to help translate the food to energy and absorb Ca and K. Magnesium improves and regulates the function of brain and prevents depression conditions. It is also essential for controlling the insulin blood levels. Taking extra magnesium is supportive in the treatment of some medical disorders like acute heart or asthma attack. It is also used in treating many lung diseases. This element has a vital role in the phosphate metabolism.

https://doi.org/10.1515/9781501516115-007

3.3 Calcium

Calcium is a macronutrient. This element is very abundant in the body. Calcium is significant for the construction of bones and teeth, for the muscle growth control and for the brain impulses. It is important in maintaining the blood pressure and in the blood clotting. It helps biomolecules digesting the food and making energy. Increased calcium consumption is needed for lowering the blood pressure, preventing some heart diseases and treating arthritis [4, 5].

Slightly soluble compounds of calcium (carbonates, sulphates and phosphates) can be incorporated into exoskeletons and endoskeletons, for example, calcium phosphate (hydroxyapatite) and calcium carbonate (calcite).

Calcium is the main constituent of the bone tissues and is approximately 1.1 kg in a normal human body. Around 10 g of this content is used for various organism functions, involving the cell function regulation, muscle contraction function, blood clotting function, as well as the enzyme regulation function. Ca^{2+} plays the role of a cofactor in hydrolases and proteins resembling Zn^{2+}. The exchange between the extracellular and intracellular amounts is realized by Ca-ATPases. Breakdown of Ca metabolism results in the deposition of slightly water-soluble compounds such as oxalate, phosphate and carbonate in the blood causing calcification.

The typical coordination numbers of Ca^{2+} in the complexes are 7 or 8, in contrast with Mg^{2+}, with a characteristic octahedral coordination. The preferred ligands are H_2O, carboxylate and carbonyl functional groups of biomolecules. Some examples are Ca^{2+} protein – parvalbumin, myofibrils, calsequestrin, calmodulin and Ca-ATPases.

The organometallics of the 2A elements are analogous to that of group 1A. They are typically ionic compounds. Beryllium, the first element of the group, behaves like lithium with more covalent carbon–metal bonds. Beryllium organometallics are extremely toxic. Dimethylberyllium decomposes and produces very poisonous BeO fumes. The magnesium organometallic chemistry is of greatest importance in the field of Grignard reagents, valuable in organic synthesis:

$$CH_3I + Mg \rightarrow CH_3Mg^+ I^-$$

Grignard reagents are toxic to the skin and pulmonary system when inhaled. They interact quickly with water and O_2. Solutions of CH_3MgBr in ethyl-ether are mostly dangerous. $Mg(CH_3)_2$ and $Mg(C_2H_5)_2$ are very chemically reactive and hazardous. $Mg(C_6H_5)_2$ is less dangerous than the simplest compounds. Contrasting the caustic group 1A oxides and hydroxides, $Mg(OH)_2$ is a moderately nontoxic compound and can be used in food additives [5–7]. Calcium, strontium and barium organometallic compounds have mostly ionic nature and high reactivity to water and oxygen. Because of their relatively difficult synthesis, their toxicological effects are of limited interest.

3.4 Role of the alkaline metals and alkaline earth metals

Biologically relevant are the group IA ions Na^+ and K^+, and the group IIA ions Mg^{2+} and Ca^{2+}. Table 3.1 represents the amounts present and daily demand of the alkaline and alkaline earth metals.

Table 3.1: Amounts present and daily demand of the metals.

	Na	K	Mg	Ca
Amounts present in man (g/70 kg weight)	105	140	35	1,050
Daily demand (in g)	1.1–3.3	2.0–5.0	0.3–0.4	0.8–1.2

There are prominent differences in the intracellular and extracellular amounts of the alkaline and alkaline earth metal cations. Table 3.2 reviews the intracellular and extracellular quantities of the vital cations and anions [5–7].

Table 3.2: Intracellular and extracellular concentrations of selected ions (mM).

	K^+	Na^+	Ca^{2+}	Mg^{2+}	Cl^-	HCO_3^-	HPO_4^{2-}	SO_4^{2-}
Intracellular in erythrocytes	92	11	0.1	2.5	–	155	190	
Intracellular value	155	110	0.001	15	8	10	65	0.5
Extracellular value	4	142	2.5	0.9	120	27	1	0

For the biological functions, mentioned above, the charge density is of fundamental significance. The charge density corresponds to the polarizing ability of the respective ion. The charge density of Mg^{2+} is very high, that is why this ion produces stable coordination compounds with N-donors, and the typical example is chlorophyll. Group IA and IIA metal ions typically do not form stable complex compounds because these ions are highly hydrated in water solutions, thus forming aqua-complexes $[M(H_2O)_x]^{n+}$.

Chapter 4
Main-group elements

Group IIIA

4.1 Boron

Boron is a micronutrient needed by the body. Minor quantities of B are required by the bones. It has a vital role in the regulation of the brain functions. According to the recent studies, small boron consumption causes reduced activity of the brain, and affects the attention and memory. Boron is necessary in the metabolism of other elements like Ca, Mg and P and for numerous physiological functions.

This element takes much consideration in the field of boron neutron capture therapy [3]. The boron distribution, when used in the cancer treatment, has been evaluated [11–14].

4.2 Aluminium

Aluminium was supposed to be unusable in the biological processes. Nowadays, it is considered to be involved in some functions of enzymes in porphyrin synthesis, such as dehydrogenase and d-amino-levulinate dehydrase. The exceeded amounts of this element from drinking water and food can be toxic though it is comparatively benign. Aluminium is used in some food additives. Recently, it was found that excess use of aluminium is associated with Alzheimer's disease. It is known that Al is partly more poisonous to plants [4, 5].

4.3 Gallium

Gallium has a similar chemical character to that of Fe^{3+} which is due to the resemblance in the ionic radii, electronegativities, ionisation potentials and electron affinities of these ions. Consequently, Ga^{3+} substances are anticipated to track the Fe^{3+} path in vivo and to occupy the Fe^{3+} centres in biological macromolecules. Subsequently, a minority of Ga(III) compounds passed in clinical tests [15]. The initial active compounds with described antineoplastic activity are chloride $GaCl_3$ and nitrate $Ga(NO_3)_3$ for oral administration. Ga(III) nitrate affects cellular metabolism of iron [16–21]. After the treatment with $Ga(NO_3)_3$, Ga affects the zinc metabolism [22, 23]. Furthermore, Ga(III) nitrate is potent against the hypercalcaemia caused by malignancies, consequently the multiple myeloma patients can be treated with this compound for bone reabsorption [24]. Despite the promising capability in lymphoma and

https://doi.org/10.1515/9781501516115-008

cancer [25], the observed nephrotoxicity restricted the application of Ga(III) nitrate. Some appropriate coordinating ligands, like 8-quinoline and 3-hydroxy-2-methyl-4H-pyran-4-one (*maltol*), have been investigated to form hydrolysis-stable coordination complexes with Ga^{3+} [26]. The orally administrated coordination compounds exhibited higher biological availability in living species and improved antiproliferative action in comparison to the respective inorganic salts [27–32].

4.4 Indium

There are some data about the usage of this element in vivo in positron emission tomography and in single-photon emission computed tomography for the diagnosis and the monitoring of patients with therapeutic interventions [33]. Despite the fact that ^{90}Y-labelled DOTA-BM couples are applied in antitumour radiotherapy, ^{111}In-labelled DOTA-BM conjugates are frequently used as imaging agents in dosimetry determination. The compound ^{111}In-TA138 was reported as an imaging surrogate for ^{90}Y-TA138 and precisely forecasted the radiation dosimetry of ^{90}Y-TA138, which has the capacity as a novel radiopharmaceutical medication for the solid tumour therapy [34].

Chapter 5
Carbon group elements

The carbon group consists of carbon, silicon, germanium, tin and lead. Excluding germanium, all the other elements and their compounds are familiar. Carbon forms a countless number of compounds in the nature. Silicon and its minerals are very abundant and they are central constituents of the Earth. Tin and lead are common in our daily life [4, 5].

5.1 Carbon

Carbon may be the most substantial element to life. The C atom is perfect to construct the biomolecules [1]. This element is a constituent of the life macromolecules, such as proteins, carbohydrates and lipids. Carbon in its elemental state is not toxic. It is the main maker of energy, as a component of sugars and fats. Its oxide CO can be readily absorbed and bound to the blood haemoglobin at very low concentrations, and that is why it is very dangerous. Being an environmental poison, it is answerable for a substantial number of intoxications. Its chronic effects of long-period exposure consist of respiratory system and heart disorders. CO interacts with oxyhaemoglobin to yield more stable carboxyhaemoglobin as given in the equation:

$$O_2Hb + CO \rightarrow COHb + O_2$$

The dioxide CO_2 is a toxicant only in comparatively great amounts. Hydrogen cyanide (HCN) along with its derivatives are notably poisonous substances with the fatal oral dose of 60–90 mg. $Na_2Fe(NO)(CN)_5$, administrated intravenously for controlling hypertension, can hydrolyse to produce cyanide that causes cyanide intoxication. Cyanide ion inhibits enzymes connected to the oxidative phosphorylation pathway [35, 36]. Carbon tetrachloride (CCl_4) and other chloro-derivatives are dangerous to the nervous system. Among hydrocarbons, the most poisonous are the halogen derivatives that contain F, Cl, Br and I, as well as S, Se, Te, N, P, As, Pb and Hg. Most of the organometallics are poisonous, while O-containing compounds of hydrocarbons are typically non-toxic [4–7].

5.2 Silicon

Elemental Si and most of its compounds are not toxic. Inhalation of SiO_2 dusts causes a serious silicosis disease. Si is found in trace concentrations in the body. In small quantities, it may be supplied in foods, though its functions and uses are not clear. Silicon is definitely a very common and abundant mineral. Its functions in growing and

https://doi.org/10.1515/9781501516115-009

maintaining the bones and connective tissues are similar to that of calcium [4, 5]. It was identified that Si decreases the efficiency of aluminium. It has been proposed that silicon can postpone or stop Alzheimer's disease; however there is no clear explanation. Generally, it is rather easy to get abundant Si in a standard nutrition and its deficiency is very rare.

Silicon, used in semiconductors, has arisen as a crucial element in the modern industry, and that is why the toxicological characteristics of Si and its compounds are of great concern. The compound of silicon that is perhaps the most toxic is silica (SiO_2) which occurs in various minerals, giving prospect for widespread human exposure [5–7]. Asbestos designates a group of silicates with formula $Mg_3P(Si_2O_5)(OH)_4$ with frequent usages. Asbestos inhalation hurts the lungs and causes a lung cancer and other pathological conditions like asbestosis, mesothelioma and bronchogenic carcinoma. Asbestos lung cancer has a strong synergetic association with the cigarette smoke exposure [37]. Compounds of hydrogen and silicon are named silanes. The simplest compounds SiH_4, SiH_2Cl_2, $SiHCl_3$ and many organic silanes and organosilicon compounds are very active, although not sufficiently known about their toxic effects [38]. The most important effects of these compounds are usually connected with the kidney damage nephrotoxicity. The silicon tetrahalides (SiX_4), where X = F, Cl, Br, I, are known to exist. Furthermore, many silicon halohydrides ($H_{4-X}SiX_X$), have been obtained and used for the synthesis of organosilanes.

5.3 Germanium

The next member of the group is germanium. The toxicology of Ge and its inorganic compounds is poorly determined. Ge is a typical trace element. Some authors trust that it is very advantageous to health. Actually, Ge possesses numerous significant medicinal assets. In the body, Ge interacts with oxygen making the body more active at receiving oxygen to the body tissues. It also helps the body excrete harmful toxins. Ge participates in some supplements for the treatment of arthritis, as well as for the therapy of allergies, high cholesterol levels and blood pressure. It is also considered as a possible anti-cancer agent. Germanium stimulates the immune system in fighting cancer cells. The most important is that germanium is non-toxic and harmless to human cells, which is not a rule for other anti-cancer treatments. Studying of new germanium chemotherapeutic anticancer agents is proceeding nowadays [1].

Corresponding to silicon, many organogermanium compounds, like tetramethylgermanium and tetraethylgermanium, are practiced in the semiconductor manufacturing to formulate deposits of this element. One of these compounds, spirogermanium, has been experienced for anti-neoplastic activity. Nevertheless, little is known about the toxic effects of organogermanium substances, though spirogermanium is of interest for anticancer chemotherapy because of its moderate toxicity.

5.4 Tin

The metal ‘tin’ is thought to be non-toxic, not harmful and is used in tin containers and in cooking tools. Organic compounds of tin, frequently used as biocides and fungicides, are poisonous. No exact biological role for tin has been recognized. Some authors suspect that very low amounts of this element are required for some animal species, for instance rats. In fact, no specific functions of tin have been recognized. The effects of excess and deficiency in tin content have not yet been proven [5].

Tin is known to have much more organometallics than all the other elements with physiologically related spectrum of activity and other applications [3]. Tributyltin (TBT) chloride and its derivatives with widespread usage have been reported to have antifungal, bactericidal and insecticidal activity. Furthermore, other TBT substances consist of naphthenate, hydroxide, bis(tributyltin) oxide and tris(tributylstannyl) phosphate [5–7]. Novel organotin(+4) sulfonamide–imine complexes with bidentate N-donor ligands were examined on various bacterial and fungal strains, and they showed higher activity than the conventional streptomycin [39]. Si(IV), Sn(IV) and Mn(II) organometal complexes compounds with tridentate imine ligands were found active against a variety of pathogenic bacterial and fungal strains [40]. Silicon(IV) and tin(IV) organometal complexes of 2-acetylfuransulfaguanidine have displayed insecticidal properties [41]. A variety of biorganotin dicarboxylates were obtained and showed anti-cancer potential [42]. Organotin(IV) ascorbates were found to demonstrate cardiovascular and anti-inflammatory activities [43]. One more Ph_3Sn(IV) thymidine derivative was reported to display moderate cardiovascular and anti-inflammatory activities. Sn (IV) organometal complex compounds with carboxylate ligand (*Z*)-3-(4-nitrophenyl)-2-phenyl-2-propenoate showed biocidal activity [44]. The tris-derivatives exhibited better activity than the biorganotin(IV) complexes [45]. Many series of organotin(IV) complexes with different organic ligands (2-maleimidoacetic acids, 1,10-phenanthroline and cephradine) displayed a range of activity against numerous bacteria and fungi [46–48]. As a rule, in a series of organotin compounds of the type R_3SnX, the observed toxicity is very high and dependent on the nature of the ligand though human contact to them does not cause many cases of intoxication [7].

5.5 Lead

Lead and its compounds are used in the production of metals just after Fe, Cu, Al and Zn. Metals usually alloyed with Pb are Sb, Ca, Sn, Ag, etc. Lead can be absorbed by the body in numerous ways. It can be inhaled in dust, accepted by food, etc. The soluble lead substances are poisonous and need special attention. The symptoms of lead intoxication include diarrhoea, constipation, vomiting, nausea, headache, dizziness and weakness. Removal of contact with Pb source is usually enough to affect a remedy. Low pollution from the usage of [$Pb(C_2H_5)_4$] has received growing consideration. Plants absorb lead from soils and traffic-induced [$Pb(C_2H_5)_4$] contamination [49]. Lead is a poisonous metal. Chronic lead toxic effects cause anaemia, cancer diseases and hormonal imbalance [50–52]. The absorption of lead is commonly by the respiratory system, especially from inhalation of lead oxides, carbonates, phosphates, halides, and sulphates. The other main path of its absorption is the gastrointestinal system and it is supposed that Pb(+2) might have the same mechanism of transportation as Ca(+2) [5–7]. Lead has a very fast transport to the bone and is accumulated there after long-period exposure. Its half-life in bones is around 20 years.

The most common lead biological effect is the inhibition of the haeme synthesis and enzymes with sulphhydryl –SH groups, though, not so greatly as Cd or Hg. The central and peripheral nervous systems are also harmfully affected by its exposure, leading to neurodegeneration, neuropathy, etc.

Some effects of lead poisoning are changeable upon elimination of the lead exposure source. Lead intoxication can be treated by using the so-called chelation therapy with EDTA, a ligand which coordinates strongly to most of metal cations. EDTA is administered in the form of Ca chelate in the lead intoxication therapy.

Organisms showed some natural protection against heavy metal intoxication. For instance, higher concentration of Ca in drinking water lowers the bioavailability of metals, especially that of Cd, Pb, Cu, Hg, Zn, and also the existence of chelates disturbs the metal uptake. Additionally, sulphhydryl groups from the intracellular metallothionein bind strongly to many heavy metals, predominantly Hg, Ag, Zn and Sn. The most examined metal for its interaction with intracellular metallothionein is Cd. Additional endogenous compounds excluding metallothionein may also participate in diminishing the toxic effects of metals and excrete them like hepatic glutathione.

5.6 Inorganic and organic lead compounds

Lead(II) inorganic compounds are principally ionic, whereas Pb(IV) compounds have a tendency to covalent binding ($Pb(C_2H_5)_4$). Lead(IV) substances, for example PbO_2, have strong oxidant properties. The basic lead salts, like $Pb(OH)_2 \cdot 2PbCO_3$, are the cause of substantial chronic intoxication. The toxic effects of organic lead compounds are predominantly remarkable as a result of the widespread usage and distribution of $Pb(C_2H_5)_4$ [53]. There is also a possibility of biomethylation of lead, and the same occurs by Hg. The toxicology of tetraethyllead was studied more widely than the toxic effects of the other lead organic compounds. This substance has a strong attraction to lipids and is highly poisonous by inhalation, absorption through the skin and ingestion. Similar to other lead organic compounds, $Pb(C_2H_5)_4$ has a strong affinity for nerve tissue and can be easily transferred to the human brain, thus reflecting on the nervous system. Retrieval from $Pb(C_2H_5)_4$ intoxication is disposed to be a very slow process. The toxicological activity of $Pb(C_2H_5)_4$ is dissimilar from the same of the other inorganic lead compounds. As a demonstration of the mentioned difference, the therapy with chelate agents appears to be unsuccessful for dealing with $Pb(C_2H_5)_4$ toxic condition. The toxic effects of $Pb(C_2H_5)_4$ seem to include its metabolic translation to its triethyl form [5–7].

Chapter 6
Nitrogen group elements

Group VA consists of nitrogen, phosphorus, arsenic, antimony and bismuth. The elements share some principal resemblances in their chemical behaviour. The proportion of nitrogen and phosphorus in the human body is very high –2.4% and 0.9%, respectively. These two elements are typically non-metals; As and Sb are metalloids and Bi is a typical metal.

The nitrogen group elements display great variability in their properties. Nitrogen is a gas at normal conditions. Phosphorus is present in various allotropic modifications, including white, red and black modifications with different stability. Arsenic occurs mostly as a metallic solid substance, but under some conditions it may exist in yellow solid form, which is more reactive. Antimony and bismuth are silver metals.

In cooperation with C, H, O and S, nitrogen and phosphorus represent the main elements involved into the living systems. These two members of VA group can be easily translated from the soil by means of plant growth, consequently they are hugely significant constituents of the plant food, especially nitrogen and phosphoric oxide.

There are some nitrogen fertilizers containing may nitrates of Na or K, NH_3, NH_4^+ salts and numerous organic compounds. Phosphorus is abounding predominantly in the form of inorganic phosphate. Arsenic, which is known for its toxic effects, is mostly beneficial in agriculture. Arsenic compounds assist in controlling the damaging insects. The last two members of the group Sb and Bi are used mainly in metallurgy for the production of alloys, because of their exceptional and necessary properties to the obtained alloys.

6.1 Nitrogen

Nitrogen is an important component of the body, existing in protein molecules. This element is plentiful in the atmosphere air and is taken from there by plants. The human's supply is getting directly from vegetable or animal food. Most of the nitrogen-organic compounds are biologically active. Nitrogen is an inert chemical element, except when inhaled under high pressure, dissolving in the blood in high concentration. These conditions allow the excess N to release through the lung [4, 5].

The so-called nitrogen fixation is a biogenic or no biogenic conversion of N_2 into its compounds. The biogenic process (60% of the overall N supply), performed by nitrogen-fixing bacteria (*Azotobacter*), cyano-bacteria (*Anabaena*) and by symbiotic bacteria (*Rhizobium*), produces NH_4^+ ions. Non-biogenic fixation, which occurs in troposphere and stratosphere, accounts only 10% and includes the next reactions:

https://doi.org/10.1515/9781501516115-010

$$N_2 \rightarrow 2N$$

$$N + O_2 \rightarrow NO + O \rightarrow NO_x$$

6.2 Toxic inorganic nitrogen compounds

Ammonia is generally used in chemical synthesis processes. The evaporation of NH_3 in liquid state can cause cryopathy. Because of its skin corrosiveness, ammonia can affect eye tissue. When breathed in, NH_3 causes bronchioles. Due to its high water solubility, ammonia can be easily absorbed by the upper respiratory system tissues causing variations of lung diseases [5–7].

Hydrazine N_2H_4 is mainly hepatotoxic and causes growth of triglyceride compounds in the liver. N_2H_4 induces hydrolytic reactions of glycogen, thus producing glucose in excessive levels and causing hyperglycaemia. N_2H_4 also inhibits enzymes' activity. The most serious toxicological result of N_2H_4 is its capability to cause methylation of DNA, which leads to cancer. In this context, N_2H_4 inhalation is connected with lung cancer conditions.

Nitrogen monoxide can be formed under tropospheric and stratospheric conditions and additional oxidation reactions to NO_2 easily occur:

$$N_2 \rightarrow 2N$$

$$N + O_2 \rightarrow NO + O$$

$$2NO + O_2 \rightarrow 2NO_2$$

At additional concentration of O_2 and H_2O, HNO_3 is produced, which is a constituent of the so-called acidic rain:

$$2NO_2 + H_2O + \frac{1}{2}O_2 \rightarrow 2HNO_3$$

Technologically, nitrogen monoxide is synthesized by transition of a mixture of NH_3 and O_2 over Pt nets at high temperature, continued to nitric acid formation:

$$2NH_3 + 2\frac{1}{2}O_2 \rightarrow 2NO + 3H_2O$$

$$2NO + 1\frac{1}{2}O_2 + H_2O \rightarrow 2HNO_3$$

HNO_3 is used for the production of NH_4NO_3 as a fertilizer.

NO is a pollutant from the automobile gases along with other constituents. It can be quickly oxidized to nitrogen dioxide. Under the UV influence, nitrogen dioxide is split into nitrogen monoxide and O atoms with high oxidizing activity to alkane- to alkyl-hydroperoxides, and form O_3 (the so-called summer smog):

$$NO_2 + h\nu \rightarrow NO + O$$

$$O + O_2 \rightarrow O_3$$

$$O + C_2H_6 + NO_2\cdot \rightarrow C_2H_5O_2H + NO\cdot$$

In stratosphere conditions, nitrogen monoxide catalyses O_3 reduction:

$$NO + O_3 \rightarrow NO_2 + O_2$$

$$NO_2 + O \rightarrow NO + O_2$$

$$O_3 + O \rightarrow 2O_2$$

Nitrogen monoxide is an important neurotransmitter in organisms [2]. Biological synthesis of nitrogen monoxide is realized by oxidation of amino groups of arginine, thus changing Arg through hydroxy arginine to citrulline [2].

Nitrogen monoxide blocks mitochondrial cytochromes through the coordination with Fe and therefore the O_2 consumption in respiration.

Higher quantities of NO are poisonous, because nitrogen monoxide binds to Fe and Cu cores of enzymes. Haemoglobin has affinity to NO which is much more effective than that of O_2. Further NO poisonousness arises from NO nitrosylation of amines, through the intermediate production of HNO_2, thus forming cancer-causing nitrosamines:

$$NO + H_2O \rightarrow HNO_2 + e^- + H^+$$

$$R_2NH + HNO_2 \rightarrow R_2N-NO + H_2O$$

Of the NO_x constituents, nitrogen dioxide is normally considered as the more poisonous, although all nitrogen oxides are highly toxic. Inhalation of nitrogen dioxide causes serious lung conditions as pulmonary oedema and bronchiolitis fibrosa. The biological action of nitrogen dioxide involves enzyme systems disturbance. NO_2 possibly acts as a typical oxidizer similar to O_3, including the production of free radicals, mainly HO·. Similar to O_3, nitrogen dioxide causes lipid peroxidation [5–7].

N_2O, identified as laughing gas, can be used as an oxidant agent and in surgery as a universal anaesthetic. It is regarded as a depressant of the central nervous system.

Nitrogen halides N_nX_x (X = F, Cl, Br, I) are considered as very poisonous compounds, mostly as eyes and skin irritants. Direct contact to nitrogen halides is limited because of their chemical activity, which destroys the respective compound before the possible exposure. NI_3 is extensively and detonating.

Halogen azido compounds XN_3 (X = F, Cl, Br, I) are enormously reactive and explosive. Their chemical interactions with H_2O can yield poisonous fumes of the respective halogen, HCl and nitrogen oxides.

Monochloramine and dichloramine are water disinfectants and are formed in the drinking water purification for providing chlorine to afford long-lasting decontamination.

Working as disinfectants, the chloramine compounds have some poisonous effects inhibiting acetylcholinesterase activity.

6.3 Phosphorus

The most abundant phosphorus is a solid. United with oxygen, phosphorus forms phosphoric acid and phosphates. Phosphorus in the form of phosphates is the most important constituent of bones and teeth. In some of its forms, phosphorus is a toxicant when taken in solid state or inhaled. The phosphorus compounds exist in fats, bones and in many proteins [4, 5].

Since P can be found in nearly all food sources, a lack of this element is hardly observed. Nevertheless, P deficiency can arise at taking some antacids for the treatment of gastritis and reflux (Mg and Al phosphates). Phosphorus is significant in maintaining the energy and blood sugar levels, which is why not getting sufficient amount of P may affect the organisms energy level. The body should maintain a strict balance between Mg, P and Ca.

In the body, phosphorus typically appears in the form of phosphates. In other forms, this element can be very poisonous. White P affects the skin and it is dangerous when ingested. Phosphine and its organic derivatives are the most poisonous phosphorus compounds. Some esters of H_3PO_4 were found to cause paralysis when ingested.

White phosphorus is a colourless waxy solid which is extremely toxic. It burns in air to yield a P_4O_{10}:

$$P_4 + 5O_2 \rightarrow P_4O_{10}$$

This modification of phosphorus is absorbed in the body through inhalation, orally and through the skin. The effects include anaemia, dysfunction of the gastrointestinal system and bone breakability. Chronic intoxication occurs at the inhalation of low amounts of white P leading to eye damages and musculoskeletal effects [5–7].

6.4 Inorganic phosphorus compounds

Phosphine, PH_3, is a colourless gas which undergoes autoexplosion at 100 °C. PH_3 is used in organophosphorus synthesis. It is a probable hazard in manufacturing processes. PH_3 gas is an irritant of the pulmonary system. Chronic effects include hepatotoxic indications, nervous system irregularities and jaundice [5–7].

Arsine, AsH_3, has many similarities with PH_3. AsH_3 is mainly produced by chemical industry of the metal refining. This is a highly poisonous compound with key effects on the red blood cells (haemolysis) [54]. Symptoms of this is the existence of haemoglobin in urine (haemoglobinurias). Acute symptoms of AsH_3 intoxication include headache, nausea, anaemia and vomiting.

Phosphorus pentoxide (P_4O_{10}) is usually produced by the burning of white P and other phosphorus compounds is P_4O_{10}.

$$P_4 + 5O_2 \rightarrow P_4O_{10}$$

The white powder of P_4O_{10} interacts with water to produce H_3PO_4:

$$P_4O_{10} + 6H_2O \rightarrow 4H_3PO_4$$

This dehydrating activity and acid formation makes P_4O_{10} a very corrosive and dangerous skin and eyes irritant.

P can form halides in both oxidation states, PX_3 and PX_5. The most stable and typical of them are PF_3 and PBr_5. PCl_5 is used in catalysis and in the synthesis of oxychloride $POCl_3$. Phosphorus halide compounds interact aggressively with water-producing oxophosphorus acids and the corresponding hydrogen halide derivatives:

$$PCl_5 + 4H_2O \rightarrow H_3PO_4 + 5HCl$$

Because of the dehydrating activity and acid formation tendency, the halides of phosphorus are very toxic strong skin and eyes irritants [5–7].

The oxyhalides of phosphorus POX_3 can be fluoride, chloride and bromide. $POCl_3$ usages are comparable to those of PCl_3, meaning, in organic and inorganic synthesis. $POCl_3$ interacts with water to produce HCl and H_3PO_3. This liquid substance gives toxic vapours, and it is a toxic strong skin and eyes irritant. Phosphorus oxychloride $POCl_3$ can be metabolized to phosphorodichloridic acid, which phosphorylates acetylcholinesterase [55].

6.5 Arsenic

In spite of arsenic's toxic status, this element may really be essential for the health. It is necessary for correct growth and reproduction. It is also supposed to be required for the nervous system functions. As it exists in food and water, which is why human beings have arsenic and a deficit in humans has not been found [4, 5].

Arsenic trioxide As_2O_3 has been accepted by the Food and Drug Administration as injection (Trisenox®) in the treatment of acute promyelocytic leukaemia with a complete remission rate between 85% and 93%, when treated with As_2O_3. The drug is under estimation for the therapy of chronic lymphocytic leukaemia, multiple melanomas and different solid tumours like neuroblastoma gastric and cervical tumours. The mechanism of action contains multiple pathways – angiogenesis inhibition, stimulation of the cell differentiation, inhibition of the cell proliferation and apoptosis induction.

The anti-oncogenic activity of arsenic and As_2O_3 is supposed to involve combined effects of dosages, distance between the acute and chronic exposure and active species (arsenate, arsenite, alkyl derivatives of arsenic acid, etc.) [56–58]. It is suggested that chronic and acute arsenic intoxication may be reversible [59]. Over the centuries, the well-known Fowler's solution containing 1% $KAsO_2$ was a common therapy for chronic myelogenous leukaemia till the current radiation and progressive chemotherapy succeeded.

The toxicological effects of As and its compounds differs broadly, starting from the exceptionally toxic arsine and the organic derivatives towards the metal As, which is comparatively not active [4, 5]. As mentioned above, arsenic substances on the whole are irritants causing dermatitis.

Nevertheless, arsenic compounds have been used as pharmacological agents, for instance anti-syphilitic agents, like Salvarsan, with a formula 3,3′-diamino-4,4′-dihydroxyarsenobenzene dihydrochloride. Arsenic compounds were used to treat amoebic dysentery and neurological trypanosomiasis, effective drug being organoarsenical Arsobal.

Inorganic arsenic compounds are known to be more toxic than its organic compounds. As_2O_3 is the most common arsenic compound in air, whereas arsenates AsO_4^{3-} or arsenites AsO_2 are water and soil pollutants. Arsenic was described to be connected with hypertension and cardiovascular impacts, diabetes mellitus effect, as well as hepatic damage [60–67]. $AsCl_3$ and the organic arsenic compounds are very harmful.

The arsenic chemistry is so wide-ranging that it is problematic to describe all of the possible effects. Its typical oxidation states are +3 and +5. Usually, compounds in +3 oxidation state (e.g. arsenite) are normally more poisonous. The transformation to As(V) is normally favoured and reduces its total hazard.

As per the biochemical studies, As acts a coagulant of proteins, producing complex compounds with coenzymes, and inhibitor of the adenosine triphosphate production. Like Cd and Hg, this element is a sulphur-seeking one. It shows chemical resemblances to P and can substitute for P in some bioprocesses, for instance, the reaction of the enzyme-catalysed 1,3-diphosphoglycerate synthesis from glyceraldehyde-3-phosphate. When AsO_3^{3-} is existing, it interacts with glyceraldehyde-3-phosphate to form a product that hydrolyses, thus preventing the formation of ATP.

6.6 Organoarsenic compounds

Two main sources of organo-arsenic compounds are possible – industrial application and biomethylation of inorganic As by different microorganisms. The reactions below demonstrate the formation of organoarsenic substances by bacteria. In reducing and acidic conditions, As(V) can be reduced to As(III):

$$H_3AsO_4 + 2H^+ + 2e^- \rightarrow H_3AsO_3 + H_2O$$

By means of methylcobalamin action in bacteria, As(III) is methylated to methyl-, and then to dimethyl arsinic acid:

$$H_3AsO_4 \rightarrow H_2(CH_3)AsO_3$$

$$H_2(CH_3)AsO_3 \rightarrow H(CH_3)_2AsO_2$$

Dimethyl arsinic acid is further reduced to volatile dimethylarsine:

$$H(CH_3)_2AsO_2 + 4H^+ + 4e^- \rightarrow H(CH_3)_2As + 2H_2O$$

$H_2(CH_3)AsO_3$ and $H(CH_3)_2AsO_2$ acids are organo-arsenic compounds that are almost certainly to be met in the atmosphere.

Although they have toxic effects, the organo-arsenic compounds have been the first pharmacological agents in 1900s, which were used to treat sleeping sickness, for instance sodium salt of 4-aminophenylarsinic acid (atoxyl). Some organoarsenic compounds that are cytotoxic were found to have anti-neoplastic activity, for example, 2-methylthio-4-(2′-phenylarsenic acid)-aminopyrimidine, an agent active against leukaemic and breast cancer cells. Roxarsone has a relatively high acute toxicity [68].

6.7 Antimony

Antimony and many of its chemical compounds are highly poisonous. Actually, the usage of Sb substances in medicine was temporarily forbidden owing to the number of fatal cases they had caused. A hydrated Na antimonyl tartrate, the so-called tartar emetic, is presently used due to its expectorant, diaphoretic and emetic properties. The maximal acceptable concentration of this element in air is near the same as for As [4, 5].

6.8 Bismuth

There are some reports on the pharmacological activity of bismuth. ^{213}Bi-DOTA-biotin displayed a healing effect, even though its low therapeutic index [69]. Some triaryl-bismuth(V) compounds of di(*N*-*p*-toluenesulphonyl)aminoacetates have exposed cytotoxic activity to some tumour cell lines (HL-60, PC-3MIE8, BGC-823 and MDA-MB-435) at low concentrations [70]. Some Bi(III) complexes of thiosemicarbazone have demonstrated selectivity and effective antibacterial activity against some grampositive bacteria [71]. Indication has appeared that the treatment by colloidal Bi subcitrate is connected with the decrease in bacterial growth and a parallel reduction in gastritis in vivo [72]. Herceptin has been radiolabelled with In-111 and Bi-213 and showed high effectiveness without any compromising immunoreactivity [73].

Chapter 7
Oxygen group elements

Group VIA or the group of chalcogens contain oxygen, sulphur, selenium, tellurium and polonium chemical elements. Oxygen is the fourth element in abundance after H, He and Ne in the universe.

Oxygen is very essential to life and to all the living organisms, taken up by animal organisms, which translate it to CO_2 and the plants sequentially use CO_2 returning O_2 to the atmosphere. Oxygen reacts with all the other elements forming different compounds and many of these processes go together with the heat and light release. The combination of O_2 with carbohydrates, fats and proteins in the body produces very slow combustion, giving heat and energy. The main supply of O_2 comes from the air, nonetheless it is accompanied by the food and water intake [4, 5].

7.1 Oxygen

It is obvious that without O_2, plants and animals cannot survive. Water, formed by H_2 and O_2, is extremely important for living bodies. Actually, over 50% of human bodies contain water, which dissolves life-supportive constituents and transfers them to cell fluids. Consequently, O_2 is a multipurpose element and most important matter to life [4, 5].

Molecular O_2 is crucial for the life processes of the aerobic organisms, though the exposure to extreme O_2 levels can lead to poisonous responses [5–7]. Even at standard O_2 levels, some poisons can cause toxical lesions. The oxidation process of the nutrient glucose is represented by the equation:

$$C_6H_{12}O_6 + 6O_2 \rightarrow 6CO_2 + 6H_2O + \text{energy}$$

In this process, molecular O_2 is the oxidizer and reduces to −2 oxidation state in H_2O. The reduction process of O_2 is multi-stepped, where active intermediates are yielded which can strongly damage the cell membranes lipids, DNA in the nuclei and proteins. At normal conditions, these reactive oxidant species (ROS) participate in further reactions or can be scavenged by antioxidants or by some enzymes action. Though, at excessive contact to oxidants and toxicants, damaging levels of ROS can be obtained. The metabolic translation of O_2 to oxygen in the O^{-2} in H_2O can be regarded as the sequential addition of e^- and H^+ cations to O_2. The first step is adding of e^- to O_2 to yield reactive superoxide ion $O_2{\cdot}^-$:

$$O_2 + e^- \rightarrow O_2{\cdot}^-$$

https://doi.org/10.1515/9781501516115-011

In the formula, the dot signifies an unpaired electron. Species that contain unpaired electrons are extremely reactive free radicals. Further, the addition of H^+ to superoxide ion $O_2\cdot^-$ produces new reactive radical, hydroperoxyl radical $HO_2\cdot$:

$$O_2\cdot^- + H^+ \rightarrow HO_2\cdot$$

Additional e^- and H^+ may interact with the hydroperoxyl radical $HO_2\cdot$, a process corresponding to the production of hydrogen peroxide:

$$HO_2\cdot + e^- + H^+ \rightarrow H_2O_2$$

H_2O_2 is produced from $O_2\cdot^-$ by the action of the enzyme superoxide dismutase. The enzyme catalase acts on H_2O_2 to form O_2 and H_2O. H_2O_2 may also be removed by the action of the enzyme glutathione peroxidase, creating the glutathione oxidized form. In the presence of suitable metal ion catalysts, H_2O_2 may undergo the Haber–Weiss reaction:

$$O_2\cdot^- + H_2O_2 + Fe^{2+} \rightarrow Fe^{3+} + O_2 + OH^- + HO\cdot$$

and the Fenton reaction:

$$H_2O_2 + Fe^{2+} \rightarrow Fe^{3+} + OH^- + HO\cdot$$

to form hydroxyl radical $HO\cdot$.

Superoxide $O_2\cdot^-$, hydroperoxyl $HO_2\cdot$ and hydroxyl $HO\cdot$ radicals accompanied by H_2O_2 affect tissues and DNA. The respective damage is denoted as oxidative lesions. Some toxic agents have the capability to endorse the formation of ROS and to oxidative stress condition, when the nucleic acids, lipids and proteins are damaged by ROS. Additional very harmful effect of ROS is lipid peroxidation. In this process, polyunsaturated fatty acids of lipids are affected and oxidized, a negative effect to lipid-rich cell membranes.

Superoxide $O_2\cdot^-$, hydroxyl $HO\cdot$ radicals and H_2O_2 are recognized as prooxidants, while the compounds that neutralize these effects are named antioxidants. Oxidative stress arises when the balance between prooxidant and antioxidant effects becomes much favourable to the prooxidants. The prooxidants effects can be deactivated by the direct interaction with small antioxidant molecules, involving glutathione, ascorbate and tocopherols. Additionally, ROS may be scavenged from the living systems by some enzymes, involving superoxide dismutase, peroxidase and catalase. Perhaps the most significant antioxidant is glutathione, which interacts with ROS to produce H_2O and the glutathione oxidized form [5–7].

7.2 Ozone

The strong oxidizing form of O_2 is ozone O_3. Ozone is the most dangerous pollutant due to its existence in the contaminated atmosphere. The interactions for the O_3 production are given below. The photochemical reaction accountable for the O_3 presence takes place in the stratosphere when O_2 absorbs UV radiation:

$$O_2 + h\nu \rightarrow O + O$$

The O atoms obtained by the dissociation process of O_2 interact with O_2 molecules to yield ozone O_3:

$$O + O_2 + M \rightarrow O_3 + M$$

where M can be N_2 molecule, which absorbs excess energy.

The obtained ozone causes the increase of temperature in the stratosphere. O_3 assists as an important filter to eliminate the dangerous ultraviolet radiation which causes skin cancer and other conditions in living creatures.

Ozone can cause pulmonary oedema. It irritates the eyes and the respiratory system. That is why the main toxicological problem with ozone includes the lungs.

Contact to O_3 rises the activity of enzymes responsible for the free-radical-scavenging activity in the lung, symptomatic of ozone's capability to generate ROS and oxidative stress. Similar to NO_2, ozone can be included in lipid peroxidation or reactions with –SH groups. Ozone seems to have contrary immunologic effects. Antioxidants and substances containing sulphhydryl groups can defend the living organisms [5–7].

7.3 Sulphur

Sulphur is a significant element which helps construct the human body. Sulphur protects the cells from pollution hazards and radiation. Accordingly, sulphur supports the liver functions, the digestive processes and so on. Sulphur is also imperative for blood clotting. In addition, sulphur is a part of insulin and vitamin B1.

In the nature, sulphur occurs in its free state as well as combined with other constituents. Sulphur exists under normal conditions as an yellow non-metallic solid substance which is not soluble in water. It interacts with all the metals except Au and Pt, producing sulphides. Sulphur can form compounds also with some non-metallic elements. Sulphur is chemically inactive and harmless, nonetheless some of its compounds are very toxic.

7.4 Inorganic compounds of sulphur

Sulphur is a constituent of some of the most poisonous compounds. Its elemental form S_8, has low toxic activity, though chronic inhalation of S_8 can irritate the membranes of mucous.

Hydrogen sulphide H_2S is produced in large amounts as a by-product of coal and petroleum industry. In the nature it can be released from volcanoes. Hydrogen sulphide is a main source of elemental sulphur by the reaction:

$$2H_2S + SO_2 \rightarrow 2H_2O + 3S$$

Hydrogen sulphide is a very poisonous compound, like hydrogen cyanide, the toxicological effects of which are greatly similar. Corresponding to cyanide, H_2S inhibits the cytochrome oxidase responsible for the respiration processes. Hydrogen sulphide disturbs the central nervous system, causing dizziness, headache and excitement [5–7]. Sulphide can also cause limited toxic effects at contact, such as pulmonary oedema, conjunctivitis and so on. There are many results of chronic H_2S intoxication.

Hydrogen sulphide acts by methylation with thiol-*S*-methyl transferase. The products are methanethiol $HSCH_3$, dimethylsulphide H_2CSCH_3 which are quite poisonous. H_2CSCH_3 is the main instable sulphur compound released from oceans and produced by the anaerobic marine microorganisms action. Bacteria produce large amounts of hydrogen sulphide and $HSCH_3$.

It has been suggested that the mucous colon membranes have some enzymes that translate hydrogen sulphide and $HSCH_3$ to harmless thiosulphate $S_2O_3^{2-}$ [74]. The nitrite-induced formation of blood methaemoglobin was used effectively to treat hydrogen sulphide intoxication. Identical to cyanide, hydrogen sulphide bonds to Fe(III) in methaemoglobin so that it is not accessible for cytochrome oxidase inhibition [75].

Sulphur dioxide SO_2, which is a common intermediate in the sulphuric acid production, is an air toxin formed by the pyrite combustion and naturally bound sulphur in fuel oil, as depicted by the next reactions:

$$4FeS_2 + 11O_2 \rightarrow 2Fe_2O_3 + 8SO_2$$

$$S\,(\text{organic}) + O_2 \rightarrow SO_2$$

Sulphur dioxide is basically responsible for the acidic rain. It is a skin, eyes, mucous membranes irritant and affects the respiratory system. it has a good water solubility and can be removed in the respiratory system [76]. Dissolved in water, sulphur dioxide produces sulphurous acid (H_2SO_3), hydrogen sulphite ion (HSO_3^-) and sulphite ion (SO_3^{2-}). The salt sodium sulphite (Na_2SO_3) was used as a food preservative [77].

Sulphuric acid participates in the production of phosphate fertilizer, and a wide assortment of important chemical compounds. Nowadays, H_2SO_4 is a main participant in acid precipitation, so it the strongest irritant taking place in acid polluted air. Most of the contaminant sulphur that develops sulphuric acid is produced to atmosphere in the form of SO_2. Sulphur dioxide emissions go together with metals, like V, Fe and Mn which catalyse the SO_2 oxidation to sulphuric acid:

$$SO_2 + 1/2\,O_2 + H_2O \rightarrow H_2SO_4$$

H_2SO_4 is a corrosive toxicant in the concentrated solutions. It easily penetrates skin and causes necrosis. H_2SO_4 fumes are eyes, lung and respiratory tract irritants. The effect of sulphuric acid is almost synergistic with sulphur dioxide.

Carbon disulphide (CS_2) is a toxic substance. The contact of skin with carbon disulphide causes skin conditions. The strongest toxic effects of CS_2 are connected with the central and peripheral nervous systems [78].

7.5 Selenium

Selenium is a very rare element, sporadically found in uncombined state, associated with natural sulphur or in amalgamation with Cu, Hg, Pb, Ag in some mineral compounds. Selenium compounds are poisonous to animals and plants. In spite of its status as a toxicant, this heavy metal is essentially very significant for human health. Se is a central part of a biomolecule that defends blood cells from harmful chemical compounds. In cooperation with vitamin E, Se helps the immune system in producing antibodies. Se is required to make the body tissues elastic [4, 5]. As with almost all the other elements, it is not difficult to take sufficient Se from a well-balanced nutrition. Superfluous quantity of Se is observed as prooxidant and is poisonous for all living species. The mechanism of action the Se toxicology is not precisely recognized. It is usually connected with oxidative stress mode of action [79].

A shortage of this element was connected to the leukaemia arthritis development and other diseases. Investigators have observed that the lower the quantity of Se in the blood, the higher the hazard of developing cancer. Actually, some scientists believe Se is an influential anticancer agent [80]. It has been suggested that Se, through the seleno-proteins, may be beneficial in stopping cancers connected with chronic inflammation processes and infections [81–83].

7.6 Organoselenium compounds

Dimethyl selenide ($(CH_3)_2Se$), dimethyl-diselenide ($(CH_3)_2Se_2$) and dimethyl selenone ($(CH_3)_2SeO_2$) can be formed by some organisms which convert inorganic Se to dimethyl selenide. Some fungi are particularly proficient at this bioprocess of methylation. The biotranslation of Se(+2) and Se(+6) to dimethyl selenide and dimethyl diselenide happens in animals. Additional organoSe compound obtained by bacteria is dimethyl selenone. The synthetic inorganic compounds of selenium are relatively poisonous, attaching the protein SH-groups, like As. On the whole, organic selenium compounds are observed as not as much poisonous in contrast of the inorganic ones.

7.7 Tellurium

The silvery-white element tellurium has chemical properties midway between the metal and non-metal properties. Comparable to Se, it cannot be found in uncombined state. It exists in the nature usually with Cu, Pb, Ag and Au, and can be obtained primarily as a by-product of the purifying of Cu and Pb. Inorganic tellurium is used in some specific alloy products. Because of its rareness in geosphere, the biomethylation of tellurium is questionable to be a main ecological problem. As a whole, the toxicity of Te substances is fewer than the same of its Se analogues. Organocompounds of Se and Te are of substantial ecological and toxicologic position. Organic compounds of selenium and tellurium can be obtained synthetically or by some microorganisms.

7.8 Polonium

Polonium is a tremendously rare radioactive chemical element which can be observed in some uranium minerals. As a basis of α-radiation, polonium has been used in some scientific experiments.

Chapter 8
Halogen elements

The members of the halogen elements group VIIA are fluorine, chlorine, bromine, iodine and astatine. Owing to their excessive activity, the free states of these elements are not observed in the nature. They are mostly found in combined forms. They demonstrate similarities regarding their chemical properties.

Iodine shortage is a serious health problem with a common symptom like the abnormal growth of the thyroid gland [4, 5]. Elemental iodine is a lethal toxicant if taken in large concentrations. Iodide salts are comparable in toxic activity to bromide salts. All the halogens are poisonous. F_2 and Cl_2 are highly corrosive and very harmful to the unprotected tissue. The elements are usually toxicologically comparable to their compounds.

8.1 Fluorine

Fluorine F_2 can be obtained from CaF_2 by delivering HF with H_2SO_4 and after the electrolysis of HF with KF in a mixture of the ratio 4:1:

$$2HF\,(\text{melted KF}) \rightarrow H_2(\text{at cathode}) + F_2(\text{at anode})$$

Of the halogens, F_2 is the most active and electronegative. In its chemical compounds, its constant oxidation state is –1. This element has many manufacturing usages. Numerous organic fluorine compounds, like chlorofluorocarbons, are used as refrigerant agents and organo-fluorine polymeric compounds, for instance, Teflon. Because of its high chemical activity, F_2 is very poisonous. It seriously affects the skin and mucous membranes of the nose and eyes.

This element is existing in low amounts in the body, in the form of bones and teeth fluorides. In its elemental form, it is a typical toxicant. Fluoride salts play a role in stopping tooth decay. Molecular F_2 is very reactive and has no biological role, but the F atom is combined in some bioactive compounds [4, 5]. F_2 contrasts from the other members of the group. Fluorine is added to toothpastes for its capability to contest cavities. Fluorides are responsible for making bones strong [1].

8.2 Chlorine

Chlorine is the best known of the halogen elements. Elemental chlorine Cl_2 can be obtained industrially for many usages, for example the manufacture of organochlorine solvents and for the disinfection of water. Chlorine is known as a toxic gas.

https://doi.org/10.1515/9781501516115-012

It has strong oxidant properties and interacts with H_2O to yield a strong oxidizing solution:

$$Cl_2 + H_2O \leftrightarrow HCl + HOCl$$

$$Cl_2 + H_2O \leftrightarrow 2HCl + \{O\}$$

where HOCl is hypochlorous acid with high oxidant properties and {O} is a nascent (freshly generated) oxygen.

Cl_2 is normally combined in the food with Na or K, making chlorides. NaCl is important to life. Throughout the body, Cl_2 is found in blood and fluids inside the cells. Accompanied by Na and K, chlorine transmits the electrical charges in the body fluids which permits nerve cells and muscles to work properly. Chlorine acts with Na and K in regulating the fluid amounts and pH in the bodies. Reasonably, Cl_2 is very significant in permitting the living organisms to digest the food and absorb the vital elements. Extreme sickness leads to a deficit of Cl_2 in the body. This can dangerously misbalance the physiological values of pH in the body [1].

8.3 Bromine

Bromine (Br_2) can be prepared from elemental Cl_2 and soluble bromides:

$$Cl_2 + 2Br^- \rightarrow 2Cl^- + Br_2$$

The main usage of Br_2 is for the synthesis of organo-bromine compounds like 1,1-dibromoethane [5–7]. Bromine inhalation or ingestion is poisonous. Corresponding to F_2 and Cl_2, it is a respiratory tract and eyes irritant because it affects the mucous membranes. Pulmonary oedema may be caused from Br_2 intoxication [5–7].

Bromine is not so required for the health, which is why no deficits have been identified and reported. Br_2 is supposed to be important in red algae and it was found in mollusc pigments [4, 5].

8.4 Iodine

Iodine I_2 is a solid crystal substance with a metallic presence. It is much more irritating than Br_2 and Cl_2, with similar effects. Contact to I_2 is restricted by the low vapour pressure, compared to Br_2 and Cl_2 [5–7].

Iodine is needed by existing organisms. Iodine is most popular for its antiseptic properties. This element keeps and regulates thyroid gland functions. Most of the I_2 in the bodies is kept in this organ. Its key role in life biology is as components of the thyroid hormones – thyroxine and triiodothyronine. Iodized salt is fortified with iodine. Iodide ions have antioxidant functions because of the reducing properties and the ability to detoxify ROS like H_2O_2 [1].

8.5 Compounds

Hydrogen halides (HX, X = F, Cl, Br, I) are comparatively poisonous.

Hydrogen fluoride (HF) can form destructive fumes at atmosphere. HF in water solution, named hydrofluoric acid, must be kept in plastic containers because it interacts with glass, producing SiF_4. Hydrofluoric acid is extremely irritative to contact tissues. The poisonous character of fluoride ions is not limited to their presence in hydrofluoric acid. These ions are poisonous in all the soluble fluorides, like NaF. At small levels, fluorides prevent tooth decay but at extreme levels, fluoride ions cause fluorosis condition [84].

Hydrogen chloride (HCl) has the utmost significant toxicology [5–7]. The water solution of HCl is called hydrochloric acid. In a concentrated solution (36% HCl) this acid is a main manufacturing chemical compound. Hydrogen chloride is not approximately as toxic as its analogue HF, though inhalation causes larynx spasms, pulmonary oedema and so on. Its vapour is disposed to dehydrate eyes and respiratory tract tissues. Hydrochloric acid is an ordinary physiological fluid in the form of a dilute stomach's solution.

Hydrogen bromide and hydrogen iodide are both typical skin and eye irritants and are dangerous for the respiratory and oral mucous membranes.

The interhalogen compounds are halogen compounds between themselves and in in some cases with oxygen. Some of them are important in the field of toxicology.

F_2 is a satisfactorily strong oxidant agent to oxidize Cl_2, Br_2 and I_2, whereas Cl_2 can oxidize Br_2 and I_2. The obtained compounds are named interhalogen ones. They exhibit dangerous chemical activity, reacting with water to produce hydrohalic acids and {O}. They are active oxidizers for organic materials. These properties give influence on their toxicity. They are predisposed to be strong corrosive irritants for the tissues. In some cases, the toxicity of these interhalogens resembles the poisonous properties of the respective elemental forms. The by-products of their chemical interactions like HF pose additional toxic hazards.

The oxides of the halogen elements are also unstable and chemically active. The most significant of these oxides is chlorine dioxide, used as a water disinfectant. Scientific surveys have suggested that chlorine dioxide and its metabolite ClO_2^- lead to the formation of methaemoglobin, cause reduction of the glucose-6-phosphate dehydrogenase activity and glutathione peroxidase activity, cause decreased levels of the antioxidant glutathione and increased levels of H_2O_2, and cause interruption of the red blood cells releasing haemoglobin (haemolysis) [85]. All these effects can suggest an overall haemotoxicity of ClO_2. Generally, the oxides of the halogens are highly active poisonous substances and their toxic effects and hazard features are analogous to the same of the interhalogen compounds.

The halogen elements form some oxoacids and the corresponding salts. The most significant are hypochlorous acid (HOCl) and hypochlorite salts:

$$Cl_2 + H_2O \leftrightarrow HCl + HOCl$$

HOCl and hypochlorite salts are used for disinfection purposes. They form active nascent {O}, and the subsequent oxidizing reaction is responsible for the toxic effects of HOCl and hypochlorite salts as eye, skin and mucous tissue irritants.

$$HClO \rightarrow H^+ + Cl^- + \{O\}$$

Perchlorates are the most oxidized salts of the chloro-oxoacids. Though they are not so poisonous, NH_4ClO_4 should be noticed as a prevailing oxidizer and active chemical compound. The toxicologic risk of perchlorates depends on the cations. As a whole, these salts must be regarded as typical skin irritants. Perchlorate ion (ClO_4^-) competes biologically with I^-. This can happen in the iodide uptake by the thyroid, directing to the thyroid hormones' biosynthesis. As a consequence, ClO_4^- can cause indications of iodine deficit.

8.6 Astatine

The long preservation of astatine may be beneficial for radionuclidic therapy [86]. Another study described the synthesis of astatine organic compounds resultant from ^{211}At and 3-astatobenzoic acid [87]. α-Emitting radionuclides ^{212}Bi and ^{211}At have been used in the anticancer therapy [88] and stable mononido-carboranyl derivatives have been synthesized as ^{211}At conjugates that can forward the excretion over the renal system [89].

Chapter 9
Metals in biological systems

Metal ions are essential for the maintenance of life and have substantial role in biological systems [1]. The alkali metal ions regulate the osmotic pressure in cells, and alkaline earth metals take a structural role in bones (Table 9.1) [90].

Sodium and potassium cations with typical coordination number 6 give complexes whose geometry is flexible and their preferred ligands are *O*-ether, hydroxyl and carboxylate. These complexes play a role in charge carrier, osmotic balance and nerve impulses.

Transition metal ions having constant oxidation states like Mg^{2+}, Ca^{2+}, Zn^{2+} and Mn^{2+} function in different ways depending on their coordination numbers, preferred ligands and structural geometry. Magnesium and calcium cations with typical coordination number 6 give complexes whose geometry is typically octahedral and their preferred ligands are *O*-carboxylate, carbonyl and phosphate. These complexes play a role in phosphate transfer and trigger reactions. Zinc has a typical coordination number 4 and its complexes have a tetrahedral geometry with many biological ligands, such as *O*-carboxylate, carbonyl, *S*-thiolate and *N*-imidazole. Zn^{2+} undergoes very fast ligand exchange, helping its role in enzyme processes. It is also totally resistant to redox changes in the values of living potentials. Manganese has a typical coordination number 6 and its complexes have octahedral and tetragonal geometries with many biological ligands, such as *O*-carboxylate, phosphate and *N*-imidazole. Manganese is a vital element in numerous enzymes. Its role appears to be important in the redox reactions in view of its different oxidation states +2, +3 and +4.

d-Elements that possess various oxidation states are good electron carriers – Fe^{2+} and Fe^{3+} function in nitrogen fixation in nitrogenases, electron transfer in oxidases, as well as in the oxygen transport – namely, Fe^{2+} and Fe^{3+} in haemoglobin and myoglobin or Cu^{+} in haemocyanin.

Transition metals, such as Cu^{2+}, Co^{2+}, Ni^{2+} and Mo^{4+} are connected with enzyme catalysis, for example superoxide dismutase or nitrogenase [91, 92].

9.1 Metal toxicity and homeostasis

Although many metal ions and their compounds are required for the survival of organisms, all of them may be poisonous if taken in large doses [1]. It is difficult to define which element is toxic. Some elements are toxic in their elemental state like white phosphorus, chlorine and mercury. Others such as C, N and O are harmless in their elemental forms. All of the elements can produce poisonous compounds except for group 8A gases which are not chemically active. Hydrogen cyanide (HCN) is an extremely toxic compound although hydrogen, carbon and nitrogen are non-toxic [5–7].

https://doi.org/10.1515/9781501516115-013

Many metals and metalloids, even those considered as essential, are poisonous to organisms if present in excess. On account of human activities, their concentrations have been largely increased, thus disturbing important biochemical processes, which is a threat for the plant and animals' health. Superfluous intake of metal ions gives uncontrolled actions; for example, oxidative damage, chronic inflammatory disease and cancer [4, 5]. For health maintenance, considerable research and advance efforts are being applied. Redox-active metals with multiple oxidation states such as Cu^+/Cu^{2+} and Fe^{2+}/Fe^{3+} are attributed to the oxidative stress. Nevertheless, outside free radicals, metals and their ions can also affect normal cells because of different interactions with biomolecules. Some elements like Pb and Hg and their compounds seem to be extremely toxic at negligible dosage levels [1], forming long-lasting compounds with organic species and accumulating in the body. Toxic elements and compounds like ethyl mercury compounds (CH_3Hg^+) and tetraethyl lead ($Pb(C_2H_5)_4$) remain a great problem in countries. The radioactive elements used to screen many diseases and to kill tumours are very important, but on the other hand, radioactivity creates mutations causing leukaemia and additional types of malignancies.

Many non-biological metals have typical toxic effects: aluminium is connected with Alzheimer's disease interacting with phosphates and cross-links proteins; cadmium is associated with renal toxicity blocking sulphhydryl groups in enzymes; mercury in the form of its lipid-soluble compounds damages the central nervous system; lead damages the peripheral nervous system, disturbs haeme synthesis and affects kidneys by replacing Ca^{2+} and Zn^{2+}.

Animals together with humans get exposed to the poisonous substances through the respiratory tract, the skin and the digestive tract. After entering the body, the metal may be transported to the gastrointestinal tract. Food is the main source of metals. That is why the nutritional influence for poisonous metal intake was widely investigated [4, 5].

Free ions of chemical elements are easily ionized and can be absorbed entirely by the organisms. Transition metal ions produce stable covalent complex compounds and typically interact as fragments of biomolecules such as enzymes, proteins and hormones. The behaviour of metal ions depends on their chemical interactions and oxidation states. The active ion directly combines with water molecules or anions to form inorganic compounds. Consequently, there is a minor possibility that the ion would interact with biological molecules to cause cytotoxic effects and other impacts. Such metals usually form complexes with amino acids, proteins, etc. The reactive forms of toxic metals are different. While all forms of Cd are toxic and need consideration, the organic compounds of Pb are poisonous and can be absorbed easily. On the other hand, inorganic arsenate substances of As(V) or As(III) are more toxic.

Elements and their compounds connected with living systems follow homeostatic norms. By homeostasis, biological systems tend to maintain stability and conditions that are optimal for survival. Interactions between deficits and extremes of chemical elements and their compounds affect homeostasis and cause diseases. The processes of homeostasis are complex within systems. There are also many illustrations of chemical elements and their compounds subject to medicinal therapeutic or diagnostic purposes which have been the theme of much recent research. Table 9.1 summarizes some of these purposes.

Table 9.1: Chemical elements and their compounds subject to medicinal therapeutic or diagnostic purposes.

Element	Product name	Active compound	Medicinal usage
Li	Camcolit	Li_2CO_3	Manic depression
N	Laughing gas	N_2O (nitrous oxide)	Anaesthetic
F		SnF_2	Tooth protectant
Mg	Magnesia	MgO	Antacid, laxative
Fe		Fe(II) fumarate, succinate	
Co	Cobalamin S	Coenzyme vitamin B_{12}	Dietary iron supplement
Zn	Calamine	ZnO	Dietary vitamin supplement
Zn		Zn undecanoate	
Br		NaBr	
Tc	Technescan PYP	^{99m}Tc-pyrophosphate	Skin ointment Anti-fungal (athlete's foot)
Sb	Triostam	NaSb(V) gluconate	Sedative Bone scanning
I		I_2	
Ba	Baridol	$BaSO_4$	Anti-leishmanial
Gd	Magnevist™	$[Gd(III)(DTPA)(H_2O)]^{2-}$	(anti-protozoal)
Pt	Cisplatin, platinol	*cis*-Platinum, *cis*-DDP	Anti-infective, disinfectant
Pt	Carboplatin	$[Pt(NH_3)_2(CBDCA)]$	X-ray contrast medium MRI contrast agent
Au	Auranofin	$Au(I)(PEt_3)$ (acetylthioglucose)	Anti-cancer agent
Bi	De-Nol	$K_3[Bi(III)(citrate)_2]$	Anti-arthritic Antacid, anti-ulcer

Chapter 10
Transition elements

The transition elements include the important 56 metals of the 107 known elements in the periodic table. They have similarity of the electronic configurations, atomic structures and chemical properties depending on their partially filled d orbitals [1–5]. Precisely, they form groups IIIB, IVB, VB, VIB, VIIB, VIIIB, IB and IIB of the periodic table. Most of the transition elements are presented below.

Among them, chromium, manganese, iron, cobalt, copper, zinc, magnesium, molybdenum and nickel are called essential trace elements. There is a predisposition for metals to give complex compounds with functional groups possessing donor properties except group lA and 2A metals. Usually, covalent chemical bonds are designed between the coordination ion and ligands, and the respective complexes can have a net charge, the so-called complex ions. In many cases, the organic bioligand can have two or more functional groups with electron-donor properties that may bond to the cation at the same time to form different complexes. The respective ligand is termed as a chelating agent.

In the organometallic compounds, the metal ion is bonded directly to carbon in the organic molecule. Unlike typical metal coordination complexes, which dissociate reversibly, in the organometallic substances, the respective organic parts are not generally stable. The neutral organometallic compounds have a tendency to have lipid solubility and this allows the facile mobility across biomembranes. Methylation is a normal process accountable for the mobility of the elements. Typical elements with methylated forms are Co, Hg, Si, P, S, Ge, As, Se, Sn, Sb, Pb and halogens [5–7]. Metal ions most commonly bond with amino acids in biomolecules. The donor functional groups most accessible for metal binding are amino-, carboxylic- and thiol (sulphhydryl) groups. The thiol group is a typical weak acid and is unionized until H^+ is displaced by the respective metal ions ($-SH \leftrightarrow -S^- + H^+$). Metals tend to accumulate in the respective target organs – the kidneys and liver, which are frequently affected by metal intoxication. Because of the prevalent opportunity for contact and their high poisonousness, some metals are principally well known for their poisonous effects [1–7].

10.1 Copper

Copper takes the third position after Fe and Zn. There are numerous copper proteins and enzymes throughout the living world isolated in both plants and animals [4, 5] like cytochrome oxidase; tyrosinases catalyse the formation of melanin. Many of these macromolecules are oxygen carriers in the bright blue blood organisms. Actually, there are analogous Fe and Cu compounds with diverse structures and functions. Copper is

https://doi.org/10.1515/9781501516115-014

important for many of the body functions – being a main constituent of the oxygen carriers of blood cells, it supports in protecting cells from toxic chemicals. It is also important for keeping elastic and flexible vessels and skin. Up-to-date research is focused into additional properties of Cu that disturb human health – anti-cancer properties and its influence to the heart disease, as well as its immunomodulating properties.

10.2 Copper-based anti-cancer agents

The usage of copper for cancer treatment dates back to early 1980 with the report by Petering [93] on the activity of copper thiosemicarbazones. More recent work has involved the use of copper complexes of carboxamidrazones [94] and carboxylates [95]. Carboxamidrazones are interesting because they have similar structures to the thiosemicarbazones which were the primary copper complexes to be reported for anti-neoplastic activity. Both classes of copper complexes have been explored by Padhye et al. The carboxamidrazone chemistry involves the use of ligands 2-acetylpyridine-pyridine-2-carboxamidrazone (appc) and 2-acetylthiophene-pyridine-2-carboxamidrazone (atpc). These ligands are readily synthesized by refluxing pyridine-2-carboamidrazide with excess 2-acetyl pyridine or 2-acetyl thiophene, respectively, in EtOH. The copper complexes of these ligands were also readily synthesized by adding equimolar amounts of appc or atpc and $CuCl_2$ dihydrate to methanol and refluxing for 1 h. The products precipitated out of solution as dark green crystalline material. The molecular structures are $[Cu(appc)Cl_2]$ and $[Cu(atpc)Cl_2]$. The anti-proliferative action of these copper complexes was experienced in vitro against murine melanoma B16F10 cells to determine if the different aryl substituents had any major effect on activity. Surprisingly, the pyridine complex is almost two times more active than the thiophene complex suggesting that the aryl substituent may play a role in increasing anti-cancer activity. The copper carboxylate chemistry involves the use of 5-amino-1-tolylimidazole-4-carboxylic acid, which can readily be obtained by the alkaline hydrolysis of ethyl-5-amino-1-tolylimidazole-4-carboxylate. The copper complex was then synthesized by reacting two equivalents of 5-amino-1-tolylimidazole-4-carboxylic acid with one equivalent of copper nitrate in methanol at pH 7. The product precipitates as the pure green solid. The anti-proliferative activity of this complex was also experienced against B16F10 cells and compared to the activity of the carboxylic acid ligand and copper nitrate alone. The complex is more active than either the ligand or copper nitrate alone [2].

There has been some recent work done trying to elucidate the mechanism of action of copper intracellularly. Zou and Somasundaram have explored reactive oxygen species (ROS) formation and DNA damage leading to p53 upregulation, respectively [96, 97]. It is believed that ROS can induce oxidative stress in cells and this leads to initiation of apoptosis [98]. These ROS are believed to induce apoptosis by interacting with intracellular macromolecules, by attacking the cell membrane, and causing DNA strand breaks. The data obtained indicates that increased levels of copper in the cell, known as copper overload, become quite toxic and lead to the production of ROS which causes a decrease in glutathione activity, cellular redox state changes and DNA damage, leading to cell death. It has been found that the tumour suppressor p53 gets activated in cells under stress especially when DNA damage occurs [99].

A review article [100] described the copper homeostasis path that can facilitate resistance to platinum-based anti-cancer medicines. There are studies which provided suggestion for a straight reaction between cisplatin and the protein hCtr1, and establishment that cisplatin and Cu have diverse significances on this process [101]. Investigating the Cu(II) complexes of actinomycin D led to the assumption that Cu(II) might contribute to the drug-induced poisonous effects [102]. Cu(II) complexes of kanamycin A have been tested on human A549 cells [103]. A Cu(II) coordination compound of the ligands 1,10-phenanthroline and L-threonine showed potent cytotoxic capacity against HL-60 and SGC-7901 cancer cells with 90% inhibition [104]. Cu(II) complexes with diethylenetriamine Schiff bases have been experienced against tumour cells and bacteria and revealed promising effects [105]. Ni(II), Cu(II) and Zn(II) coordination compounds of tris-amino-functionalized 1,3,5-triaminocyclohexane and meso-(4-aminophenyl)triphenylporphyrin have displayed dissimilar levels of cytotoxicity to cancer cell lines [106–107]. Copper complexes of carboxymethyl-L-methionine-*N*9 -8-quinolylamide exhibited anti-leukaemia effects [108]. Along with the anti-cancer properties of copper(II) complexes, their antibacterial and anti-fungal activities have been proven by many authors. Copper(II) thiabendazole salts showed potent antifungal and cytotoxic activities [109]. Dimeric copper complexes of the pyridoxalthiosemicarbazone derivatives have been established in vitro on several tumour cells [110]. Cu(II) complexes of N(4)-substituted derivatives of thiosemicarbazones and polypyridyls were found to have substantial activity against some bacteria [111]. Cu(II) and Ni(II) complexes of 1,3-bis(thiosemicarbazonato) and pyrazoline proligands have been tested in vitro for their cell growth inhibitory effect and apoptosis induction on U937 cell line [112, 113]. Copper(II) complex compounds of Indomethacin [IndoH] showed importantly condensed gastrointestinal toxic effects and the complexes were found less poisonous than the corresponding Zn–Indo complexes [114]. Structural description of Zn(II) and Cu(II) coordination compounds of clioquinol (8-hydroxyquinoline derivative) with activity in the treatment of Alzheimer's disease was carried out [115, 116]. The effects of copper complexes of iso- and thio-nicotinate were experienced on several bacteria and fungi and showed good activity [117]. Copper(II) complexes of tetraazacyclotetradecane derivatives have shown good antibacterial activity [118]. Co(II), Cu(II), Ni(II) and Zn(II) complexes of bis-(1,1′-disubstituted ferrocenyl) carbohydrazone or thiocarbohydrazone ligands have been tested for their antibacterial and anti-fungal activities, which have been improved because of the coordination with the metals [119]. Anti-fungal and antibacterial activities were reported for Cu(II) coordination compounds with *N*,*N*,*N*′,*N*″,*N*″-pentamethyl diethylenetriamine (PMDT) and polypyridine ligands [120]. ^{64}Cu-labeled macromolecular complexes with bi-functional chelators, have been reported as another possibility for labelling Cu radionuclides to biomolecules in radiotherapy [121]. Copper(II) complexes with tetraazamacrocyclic organic ligands and tetradecane were found optimum candidates in the field of radiopharmacy [122].

10.3 Iron

Iron is known to have numerous functions in the living organisms. Certain biomolecules in human brain are connected with the iron concentrations. Iron is significant as an immunomodulating agent. There are some definite types of iron macromolecules, for instance ferritin, haemoglobin and ferredoxins. Iron is an important part of haemoglobin – the fragment of blood that carries oxygen. The haemoglobin molecule transports O_2 to the muscles' tissues, transferring to myoglobin. An oxygen molecule covalently bonds to each of the four iron atoms in a haemoglobin molecule. CO is extremely poisonous to mammals because the carbonyl bonds very powerfully to Fe of the haemoglobin, stopping it from carrying O_2 molecules. Plants and bacteria use a group of sulphur structures with Fe(III), ferredoxins, which act as excellent electron transfer agents. Iron deficiency anaemia is the most common dietary disease [1].

Iron plays a central role in biological events due to its general availability and its specific chemical and biological properties: ease of modification between the +II and +III oxidation states; formation of hexaaqua cations in water – Brønsted acids; predisposition to form oligomers and polymers; easy transformation between low- and high-spin states; flexibility of the coordination numbers and coordination geometry.

Iron participates in vital biological functions [5–7], and consequently the deficiency of iron can be a significant target for growing of cancer. It has been observed that iron restriction by the dietary supplement evidently decreases tumour growth.

Iron typically exists in the ferric form when taken up with the food, and consequently goes to the gastrointestinal tract as Fe(III). In the small intestines, this ferric form of iron is transformed to the ferrous form 'Fe(II)'. Iron can be absorbed only in this oxidation state by the mucosal epithelium cells. For transmission to the blood, re-oxidation of $Fe^{2+} \rightarrow Fe^{3+}$ is needed.

10.4 Iron-based anti-cancer agents

There were a variety of descriptions on the anti-cancer activity of iron-chelated metal complexes [123–126]. Ferricenium picrate and trichloroacetate were the first two iron compounds exhibiting anti-tumour activity against some tumour cells [5–7]. The mechanism is through the inter-translation between non-active ferrocene(II) form and active ferrocenium(III) form of ions in hypoxic tumour cell lines.

Recent work in the field of anti-cancer activity of iron complex compounds has been explored by Richardson [127] and Kawakami [128]. The activity of iron complexes is believed to involve either in vivo redox chemistry or the targeting of superoxide dismutase (SOD). The work done by Richardson et al. involves the chelation of Fe(II) and Fe(III) to 2-benzoyl and 3-nitrobenzoylpyridine thiosemicarbazones where the iron atom is wedged between two ligands. The iron thiosemicarbazone metal compounds were all obtained by adding one equivalent of the appropriate ligand and Et_3N to EtOH followed by the addition of 0.5 equivalents of $Fe(ClO_4)_2$ or $Fe(ClO_4)_3$. The reactions were refluxed under nitrogen for 30 min and upon cooling the green products precipitated. The anti-cancer action of the iron complexes was in vitro experienced against SK-N-MC neuroepithelioma cell line. The cell lines were incubated with the respective coordination complex for 72 h and the IC_{50} concentrations are reported. All of the iron complexes were extremely active. The iron complex that was explored by Kawakami involves the complexation of Fe(III) to the porphyrin. This metal porphyrin complex was readily synthesized. The Fe(III) porphyrin complexes and cisplatin were tested in vitro against the A549, MKN28, MCF7, SW948, HeLa and PC9 cell lines. Impressively, the iron complex was more effective than cisplatin against every cell line except the breast cancer cell line. Though iron has been revealed to kill cancer cell lines in vitro, its use as an anti-cancer drug would appear not to be of utmost importance. This is due to the idea that many cancers require elevated levels of iron [129]. It is believed that iron not only enhances tumour growth but also plays a major role as an initiator to tumorigenesis. It has been shown in mice that a diet high in iron increases colon cancer tumorigenesis [130]. Therefore, it is believed that iron chelators may be better candidates as anti-cancer drugs than iron complexes. A review article describes iron oxide nanoparticles as contrast agents because of the exceptional spatial resolution and the capability to track labelled cells during continued time [131]. Functionalized iron oxide nanoparticles, possessing superparamagnetic properties, also presented benefits for intracellular imaging with negligible perturbation [132]. Some chelate ligands including 2,3-dihydroxypyridonate and 2,3-dihydroxyterephthalamide parts for Fe^{3+} have been investigated as contrast agents in MRI medicine [133]. Fe(III) ions produce a low-spin complex with the anti-tumour agent 'bleomycin' whose cytotoxic activity depended on the capability to cleave DNA [134, 135]. The more stable Co(III) compound HOO–Co(III)bleomycin was also considered as a chemical referent of Fe(III) complex of bleomycin [136].

10.5 Vanadium

Vanadium plays a vital role in some marine organisms as well as in human health. It is included in helping the organisms translate the food into energy. It was proposed that vanadium stabilizes blood sugar levels in diabetics. Vanadium is correspondingly supposed to help the formation of bones and teeth. This element is used in the treatment of diabetes, high cholesterol, low blood sugar, tuberculosis, heart and other diseases. No definite indications of deficiency have been recognized.

Some investigations have revealed that vanadium might reduce the tumour growth and offer anti-cancer protection. Its exact role in human health is still unknown, and more research is needed for its determination [4, 5].

Vanadium exists in different oxidation states and among them +5 is the most abundant. The toxicity of vanadium increases with its valence state. Even when this element exists in +4 or +5 oxidation states (as it is at physiological conditions), it can still act in the form of cations or anions such as VO_4^{3-}, HVO_4^{2-} and $H_2VO_4^-$. The inorganic compound V_2O_5 is the most toxic of the vanadium compounds. Usually, the pulmonary system suffers from vanadium toxic activity [5–7]. The most common pathological effects of contact with vanadium are bronchitis, bronchial pneumonia, skin and eye irritation. The kidneys, gastrointestinal tract and nervous system can also be affected. It was detected that vanadium possesses similar to insulin effects on the key insulin-targeted organs such as muscles and liver. It has been reported that organically chelated forms of vanadium are more active in treating the symptoms of diabetes and they are less poisonous than inorganic forms of this element [137, 138].

10.6 Vanadium-based diabetes drugs

The similar to insulin effects of some vanadium salts, peroxovanadium(V) and coordination vanadium compounds have been reported in the literature. These vanadium compounds could lower blood sugars from high to normal levels. They also possess capability to reduce cholesterol levels and blood pressure in human studies with low doses. The anti-diabetic properties are reported on animal organisms or humans with either insulin-dependent diabetes mellitus (IDDM, type I or DM I) or non-insulin-dependent diabetes (NIDDM, type II). It has been proved that vanadium compounds seem to improve the effects of insulin in organisms with NIDDM, reducing the levels of glucose, lipids and cholesterol. Vanadium substances have also been revealed to be beneficial in patients with IDDM, where they improve sensitivity to insulin. Oral therapies currently exist for diabetic treatment; however, these medications do not mimic insulin signalling. Vanadyl(IV) and Zn(II) ions and the respective coordination compounds have been observed to be useful in the treatment of diabetes in animals and in humans. The compounds consist of dioxo-(semicarbazone)vanadium(V) complexes *cis*-VO_2L which were experienced as possible insulin-mimetic agents [139]. For testing of insulin-mimetic activity, the enzymes phosphotyrosine phosphatases (PTPase) and tyrosine kinases were proven to participate in the binding of insulin receptors. The same were found to be sensitive to stimulation and inhibition of vanadium and are known as markers [1].

Mainly, V(IV) and V(V) complex compounds were tested as insulin mimetics in the form of inorganic salts or coordination complexes with organic ligands. The vanadyl ion, VO^{2+}, has been administered as vanadyl sulphate ($VOSO_4$). This inorganic vanadium salt has been the focus of clinical tests although it can have gastrointestinal side effects. Bidentate chelating compound obtained by interaction of maltol with $VOSO_4$ to produce bis(maltolato) oxovanadium(IV), abbreviated BMOV, has been obtained to improve solubility effects, capability for insulin-mimetic action and eventual oral administration [140]. The structure of BMOV is square-pyramidal for the central V(IV) ion having one unpaired electron [141]. Comparison of investigation between inorganic vanadium compounds and vanadium complexes with organic ligands has been reported [142]. Inorganic and organic compounds $VOSO_4$ and BMOV interact with HSA and apoHTf, producing different V-protein complexes. Bis(acetylacetonate)oxovanadium (IV) complexes $VO(acac)_2$, $VO(Et\text{-}acac)_2$ and $VO(Me\text{-}acac)_2$ have been reported to have continuing in vivo insulin-mimetic activity in commonly studied animal model streptozotocin-induced Wistar rats [143]. The above-mentioned vanadium complexes with organic ligands, $VO(acac)_2$, $VO(Et\text{-}acac)_2$ and BMOV, have been compared to $VOSO_4$ in oral treatment. All the complexes produced a quicker and greater reduction in glycaemia, with $VO(acac)_2$ being the most effective [144].

Peroxovanadates were experienced for their insulin-mimetic activity and revealed to stimulate insulin receptor tyrosinase kinase (IRTK or IRK) and to inhibit the activity of PTPase [145]. These V(V) compounds were found to be effective inhibitors of PTPases. Even though peroxovanadates in the solid state are stable, they are instable in water solution. Since the oral treatment is not probable, the substances were used intravenously [146, 147]. The heptacoordinate V(V) compound potassium oxodiperoxo(1,10-phenanthroline) vanadate(V) trihydrate, $[bpV(phen)]^-$, was described [148]. A monoperoxovanadate(V) complex, oxoperoxopicolinatovanadium(V) dihydrate [mpV(pic)], was reported to realize a reduction in glucose in STZ-diabetic rats by intraperitoneal or subcutaneous treatment [149].

Some vanadium(III) coordination compounds were estimated for possible insulin-mimetic properties [150]. Typically, these complexes undergo fast oxidation to IV and V oxidation states at $pH > 3$. In this study, coordination compounds with maltol (ma), ethyl maltol, kojic acid and other organic ligands were experienced for their stability in hydrolysis and oxidation–reduction reactions. The maltol complex $V(ma)_3$ was observed to be stable in hydrolysis and oxidation reactions. Its behaviour was comparable to that observed for BMOV with a formula $VO(ma)_2$. Thus, V(III) complexes can be similar to V(IV) concerning the insulin-enhancing ability.

Vanadium possesses important medicinal properties [151–154]. It has been reported to inhibit the terminal differentiation of murine erythroleukaemia and after integration in diet it decreases the possibility of chemically induced mammary carcinoma. In addition to this activity, vanadium has been revealed to reduce the growth of human prostate cancer cells and to decrease bone and liver cancer in animals. These extensive properties on tumour and diabetes suggest that this element is expected to come under more examination for possible health benefits and toxic activity. Not all of vanadium's properties are optimistic. Vanadium is involved in blocking of enzymes like ribonucleases, mutases, kinases and synthases, and this action has negative consequences [1]. A recent study investigated the cytotoxic effects of vanadium compounds, [VO(acac)$_2$] and vanadate [155]. Vanadium coordination complexes have also been developed as effective anti-cancer agents [156, 157]. The most capable multi-targeted vanadium compound with anti-cancer activity between some bis (cyclopentadienyl)vanadium(IV) and oxidovanadium(IV) compounds is bis(4,7-dimethyl-1,10-phenanthroline)sulphatooxidovanadium(IV) (metvan) [158–160]. At nano- and micro-molar concentrations, metvan induces apoptosis in human leukaemia and myeloma cells, as well as solid tumour cells resulting from ovarian, breast, testicular and prostate cancers. It is extremely active against cisplatin-resistant testicular and ovarian cancer cells. Metvan-induced apoptosis is connected with a loss of mitochondrial trans-membrane potential, the ROS production and reduction of glutathione [161]. Metvan exhibits substantial anti-neoplastic activity, delays tumour growth and extends survival time. The broad-spectrum anti-neoplastic activity of metvan together with the lack of toxicity permits further progress of this compound as a potent anti-neoplastic agent as an alternative to Pt-based anti-cancer compounds [162, 163].

10.7 Manganese

Manganese is a very important element involved in many processes in the body. Like the other transition metals, the living role of Mn appears to be as a typical redox agent, changing between its +2 and +4 oxidation states.

Manganese is a vital element in numerous enzymes of plants and animals. It also helps activate many enzymes in metabolism, regulates the immune system and blood sugar levels, and is connected with the reproduction of cells. Manganese is also significant for bone development. Moreover, it acts with vitamin K to regulate blood clotting. Manganese complements vitamins B to help control the stress effects [4, 5]. It is supposed that not receiving sufficient doses can cause reduced bone construction, and affect the fertility and the blood clotting ability. There is a relation between reduced manganese intake and higher skin cancer levels. Humans can easily get adequate manganese intake from a good composed diet [4, 5]. In mammals, manganese participates in arginase, which converts N-containing wastes to urea. There is a set of plant enzymes – the phosphotransferases, which incorporate manganese.

Manganese is the coordination ion in SOD, and several cancer cells display lower SOD2 concentrations [164, 165]. Manganese SOD (MnSOD; encoded by the SOD2 gene) is the major ROS detoxifying enzyme of cells because of its localization to mitochondria. It is a participant of mitochondrial Fe–Mn containing family of SODs [166, 167]. In human prostatic tumour cell lines, transformation of cDNA upregulates the SOD2 by sixfold, and this advancement is enough for tumour decrease. This property proposes the SOD2 as a tumour suppressor gene. An effective mimic of SOD is the compound Mn-salen (EUK-135) that displays effectiveness in the cell survival [168, 169].

The usage of Mn compounds in photodynamic therapy has been reported in the literature. The tris(pyrazolyl)methane derivative $[Mn(tpm)(CO)_3](PF_6)$ gets stimulated after UV-light irradiation, and the CO groups are free in water buffer. The compound shows photoinduced action in HT29 colon cancer cell line with an efficiency comparable to 5-FU. This specific compound as well as some of its derivatives may be possible drugs with good specificity and efficacy [170].

The Mn(II) ion has numerous positive possessions that lead to its possible use as an MRI contrast agent, especially its long electronic relaxation time, high spin number and kinetic instability leading to strong water exchange. The existing position of Mn(II)-enhanced MRI of the heart, suggesting that Mn(2+) uptake may be understood in relation to the heart function, has been reported in a review article [171]. Another research paper reported the estimation of new Mn(II)–EDTA derivatives [172]. The manganese complexes of the type $[Mn(N_4MacLn)(NO_3)_2]$ (N_4MacLn is a tetra-azamacrocyclic ligand) and the respective ligands were tested in vitro against numerous bacteria and fungi to evaluate their inhibiting activity [173].

10.8 Chromium

The transition metal chromium is an essential element in humans. It is a vital mineral that appears to have a valuable role in the regulation of insulin action to stabilize blood glucose levels. It helps the bodies captivate energy from the food and can help rise muscle mass while decreasing fat mass in the body. Chromium helps heart muscle cells absorb the necessary energy. Chromium is a component in many multivitamins and supplements; nevertheless, the body absorbs Cr well from food [1]. A deficiency of chromium(III) or an incapability to use chromium ions can cause diabetes.

In its chemical compounds, chromium is present in $+2 \div +6$ oxidation states, but the most stable are +3 and +6. In acidic medium, Cr(III) is in the form of aqua complex $Cr(H_2O)_6^{3+}$. At $pH \geq 4$, Cr(III) has a strong predisposition to precipitate in the form of hydroxide $Cr(OH)_3$:

$$Cr(H_2O)_6^{3+} \rightarrow Cr(OH)_3 + 3H^+ + 3H_2O$$

The main forms of Cr(VI) are chromate CrO_4^{2-} (yellow) and dichromate $Cr_2O_7^{2-}$ (orange) in solution. The dichromate prevails in acidic medium. The equilibrium of the reaction is forced to the left by high concentrations of H^+:

$$Cr_2O_7^{2-} + H_2O \leftrightarrow 2HCrO_4^- \leftrightarrow 2H^+ + 2CrO_4^{2-}$$

Both CrO_4^{2-} and $Cr_2O_7^{2-}$ are strong oxidizing agents at low values of pH.

Chromium in +3 oxidation state is important for the metabolism of insulin, sugar and lipids. Its deficiency gives indications of diabetes. Nevertheless, Cr is known as a toxicant. While chromium in elemental state and Cr(III) ions are non-poisonous, Cr(VI) is toxic and cancer-causing [174]. Contact to hexavalent chromium typically includes sodium chromate. The chromate salts are usually soluble in water solutions and thus easily absorbed into the bloodstream.

Chromates have been known to be potent inducers of cancer in exposed workers. Contact with atmospheric chromate may be a reason for bronchogenic carcinoma. In the body, Cr(VI) can be easily reduced to Cr(III), as depicted in the equation, though the reverse process is not possible in the body:

$$CrO_4^{2-} + 8H^+ + 3e^- \rightarrow Cr^{3+} + 4H_2O$$

The biological role of Cr(III) is connected with the regulation of the metabolism of carbohydrates and lipids, although the respective mechanism of action is still unknown [175]. There are some problems in defining the chromium essentialness: firstly, chromium(III) is passive in reactions of substitution; secondly, its concentration is very small and it is difficult to find procedures for its determination [176].

The low-molecular-weight substance (LMWCr, also known as chromodulin), containing chromium, is an oligopeptide that appears to transport chromium in the body. Chromodulin improves the insulin properties and activates PTPases to some extent and stimulates the action of IRTK [177–179]. Chromodulin exists in apo-form in the nucleus of insulin-sensitive cells. Investigations of apo-chromodulin and similar compounds of biologically active metals showed the importance of Cr(III).

A study reported that long-period contact to chromium can recover the sensitivity of insulin [180]. Substantial anti-diabetic activity of $CrCl_3$ has been revealed by in vivo trials [181]. It is known that Cr, Ni and Mn are main metal toxicants in welding gases and it has been shown that Cr(VI) and the ions of manganese seem more poisonous by testing lung cell lines in vitro [182]. Investigations on the synthetic, functional biomimetic trinuclear cation $[Cr_3O(O_2CCH_2CH_3)_6(H_2O)_3]^+$ showed relation between the insulin resistance and the cancer occurrence [183].

10.9 Cobalt

Cobalt is the metal at the core of the molecule of vitamin B-12 (cobalamin). This coenzyme has a crucial role in the nucleic acid metabolism and is needed to form the red blood cells. A related molecule, methylcobalamin, is used by some anaerobic bacteria to yield methane. Vitamin B-12 and vitamin C are water soluble and are essential for the food digestion. Furthermore, vitamin B-12 prevents nerve damage and is important for normal nerve functioning. A lack of vitamin B-12 can cause anaemia, neuropathy and other neurological consequences. It can be stored in substantial amounts in the body, and that is why it is easy to get sufficiently of this significant vitamin in the diet. Deficiencies of vitamin B-12 are usually rare in young people, but sporadically occur in adults especially in strict vegetarians because vegetables do not contain this significant vitamin [1].

Cobalt intoxication is not common. People can get cobalt intoxication by breathing in too much cobalt, swallowing too much of it or having continued skin contact with it. The extreme levels are damaging. Utmost cases of human contact to poisonous cobalt levels have been observed in workers occupied with solid cobalt alloys and tungsten carbide, from which very fine alloy particles were breathed in. The side effects of metal inhalation were observed on the lungs, characterized by pneumonia and asthmatic reactions. Contact to cobalt is similarly probable through the food and water [5–7].

Cobalt compounds were recognized for their perfect mimic of metalloenzymes. Many cobalt compounds have been widely considered for the creation of anti-tumour, hypoxic-selective and DNA-cleaving agents [184–191]. In solid tumours, some cells are distant from the blood vessel; thus, it is problematic to influence them by standard chemotherapeutic agents. The redox couple Co(II)–Co(III) can offer selectivity in hypoxic cells. The ligands have important results on the activity of these substances. [1,6-Bis(2-hydroxyphenyl)-3,4-diaryl-2,5-diazahexa-1,5-diene] cobalt(II) coordination compounds were synthesized and experienced for in vitro anti-tumour activity on different cancer cells [192]. A range of transition metal Cd (II), Sn(II), Co(II), Fe(III) and Pb(II) complexes of Schiff bases were tested, and the Cd(II) and Co(II) coordination compounds were observed to be very effective against CEM-SS and HeLa cell lines [193]. The complex Co(salen) [where salen = *N*, *N*-bis(salicylaldehyde)ethylenediimine] has been tested for anti-tumour activity, and the structure–activity relationship was constructed on the ligands used at the ethylenediamine moiety. Recent progress in this field counts the Co(III) complex of the MMP inhibitor Marimastat as an anti-metastatic agent, which displays an advanced level of tumour growing inhibition [194]. Another chemical group of cobalt substances used in anti-cancer treatment are bi-nuclear carbonyl compounds of cobalt [195–196]. The most potent substance was Co-ASS – the complex of cobalt and carbonyl with aspirin which was found effective against some cancer cells, especially breast cancer cells [197–198].

10.10 Nickel

Of the period of 3*d* transition metals, nickel's biochemistry is the most unclear. This element is a micronutrient important for many vital functions increasing the hormonal activity and participating in lipid metabolism. Nickel ions are present in some enzyme systems in the form of porphyrin-type complexes. Several enzymes (urease, NiFe hydrogenases, dehydrogenase, etc.) and one cofactor (nickel-tetrapyrrole coenzyme, cofactor F430) have been shown to contain nickel. The low levels of nickel cause liver and kidney diseases. Correspondingly, excess amounts of nickel are connected with the heart and thyroid diseases as well as with cancer. Therefore, nickel is both essential and toxic for animals and humans, and the implication of the nickel amount in the living organisms remains unknown. The main source of contact with nickel is the food because nickel is important for the plants. Nickel can be found in food and water in the nature.

Toxicologically, nickel-containing compounds are known as human carcinogens because of the high respiratory risks of cancer detected by epidemiological research of sulphide ores. The exposed people may display a skin allergy to nickel, the so-called contact dermatitis – the other key poisonous effect of nickel metal. It practically permanently occurs as the consequence of contact of nickel with the skin when wearing nickel jewellery. The carbonyl complex, $Ni(CO)_4$, is considered as a very poisonous nickel complex [5–7].

Nickel is recognized as an important trace element for plants and animals. Many beneficial effects of Ni on plant growth have been reported. Experiments show that chickens and rats develop liver problems when fed with a diet that lacks nickel. Some bacteria make specific enzymes using nickel but the function has not been defined, for instance the bacteria which reduce CO_2 to methane also use nickel. The necessity for nickel has been clarified when it was observed that most hydrogenases contain Ni in cooperation with Fe–sulphur clusters. Even though in inorganic chemistry the +III oxidation state is infrequent, Ni^{3+} ions are connected with the enzyme redox processes. There are some plants that accumulate nickel, for instance some tropical trees.

It is known that the contact to numerous nickel-containing compounds raises the hazards of lung, nasal and other types of cancers. Some authors have studied Ni complexes as potential anti-cancer agents. A nickel coordination complex with *ortho*-naphthaquinone-thiosemicarbazone has been experienced on MCF7 tumour cell line, displaying low IC_{50} values and suggesting a mode of action which includes embarrassment of the activity of topoisomerase II [199]. Ni(II) and Cu(II) complexes with the ligand *N*-methylisatin-3-picolinoylhydrazone were found potent in preventing the in vitro cell production on TOM1 and NB4 leukaemic cells [200]. Therefore, synthesis of novel anti-cancer drugs continues to be a potential topic for research.

10.11 Molybdenum

Molybdenum is the most physiologically significant element of the group 6B. This heaviest member of the group has varied functions in the body. It has several oxidation states like +4, +5 and +6. The redox potentials of these ions overlap with the physiological structures. Molybdenum is a required trace element in the diets, just like iron and magnesium, but its function and the smallest levels have not been recognized. Though human body only needs very small amounts, it is a key constituent of many vital functions. Molybdenum has a tendency to have the biggest concentration in the liver, skin, kidneys and bones. It is needed for the correct function of some substances in the living organisms. Some of them have significant role and are connected with the functions of iron and nitrogen in the human diets. This element is supposed to be significant in helping the production of human cells. Correspondingly, small quantities of it were attributed with the teeth. There are some suggestions that this element reduces the hazard of some kinds of asthma occurrences [4, 5].

It is interesting why such rare metal is so physiologically important. The ion $[MoO_4]^{2-}$ has a high-water solubility at neutral values of pH, making this ion easily transferrable in physiological conditions. $[MoO_4]^{2-}$ ion with a negative charge is more appropriate for diverse surroundings than are the 3d transition metal cations. Actually, it is claimed that the transport of molybdate ion has the same mode of action as SO_4^{2-}, which is a typical example of the resemblances of ions of groups 6A and 6B.

Situations in which too much or very slight intake of molybdenum are rare, but both have been connected with serious adverse effects. A deficiency of molybdenum can contribute to cancer diseases and anaemia. Along with the food sources, hard tap water is also a source of molybdenum to the diet [4, 5].

Molybdenum is stored mainly in the liver and kidneys, but the most amount is transformed into a molybdenum cofactor. The molybdenum cofactor activates some essential enzymes (aldehyde oxidase, sulphite oxidase, xanthine oxidase and mitochondrial amidoxime reducing component), which determine many living reactions. The absence of molybdenum cofactor is a very rare genetic disorder in which babies are born without the capacity to make molybdenum cofactor, although this condition is extremely rare.

Various enzymes depend on the content of molybdenum, which is typically absorbed in the form of $[MoO_4]^{2-}$. The most vital enzyme, containing molybdenum iron and molybdenum, is nitrogenase. This group of enzymes befalls in bacteria which reduces atmospheric N_2 to NH_3, which is useful for the plants in the synthesis of protein. Nitrogen fixation is connected with photosynthesis. The process described by N_2 fixation is the reduction of N_2 to NH_3, useful for the synthesis of amino and nucleic acids, as well as other important compounds. One interesting aspect of nitrogen fixation is the role of molybdenum in the enzyme nitrogenase [1]. All Mo-containing nitrogenases contain two parts: the Fe-protein and the MoFe-protein. The Fe and Mo atoms are situated in metal–sulphur clusters [201]. Obviously, the N_2-fixing bacteria uses enzymes that hold both Mo and Fe. Many investigators have studied different models for probable intermediates in these reactions [202–204]. Lehnert and Tuczek studied coordination potential by using DFT calculations on the compounds $[Mo(N_2)_2(dppe)_2]$, $[MoF(NNH)(dppe)_2]$ and $[MoF(NNH_2)(dppe)_2]^+$, with the ligand dppe = 1,2-bis(diphenylphosphino) ethane [205]. There is another group of Mo-containing enzyme structures, the so-called molybdopterins, which have a central moiety containing MoS_2 as well as an organic moiety identified as a pterin. The same enzymes frequently contain additional metal, predominantly an Fe–S structures, and they play the important role in oxidation–reduction reactions.

10.12 Tungsten

Tungsten has a minor function in life processes. It is not identified if humans need tungsten. Nevertheless, this element has been the subject of many in vivo experimental and in vitro studies in order to regulate its metabolism and toxicity. Opinions are different regarding the need for tungsten in biological processes, even if it has been evidenced to be needful for some bacteria. Tungsten and its compounds are not considered very toxic for humans.

Tungsten is supposed to be used by a minor group of enzymes in a mode comparable to that of Mo [4, 5]. W-containing enzymes occur in some non-oxygen bacteria in hot ocean environments, the hyperthermal Archaea. Tungsten have compounds in +4, +5 and +6 oxidation states. As the above-mentioned bacteria occur at high temperatures, it is claimed that W rather than Mo is used by the enzymes because W has a strong metal-to-ligand bond, permitting the enzyme to work at high temperatures without decomposition. Therefore, the reaction rate for the W enzyme at around 110 °C is analogous to that of the Mo enzyme at physiological 37 °C. Precisely, how W is utilized by these bacteria is relatively unclear. Some molybdoenzymes can retain function when molybdenum is replaced with tungsten, and some organisms have both tungsten-dependent and molybdenum-dependent isoenzymes that are expressed under different environmental conditions.

10.13 Titanium

Very little is known about the physiological role of titanium. This element has no identified physiological usage in humans, though it is known to stimulate some processes. It is non-toxic even in large doses due to its resistance to corrosion from bodily fluids. That is why Ti is used in prosthetics. In some plants, Ti is utilized in the energy production.

10.14 Titanium as an anti-cancer agent

Titanium complexes represent good alternatives to Pt-based anti-neoplastic drugs because of their low toxic effects, due to the hydrolysis of Ti(IV) coordination compounds in biological water-based location to the inert titanium dioxide. Ti(IV) complexes were the first to enter clinical trials in anti-cancer treatment following the accomplishment of Pt-based chemotherapy, with the original complexes titanocene dichloride and budotitane. Many derivatives with different labile organic ligands or substituents contributed to improve the anti-cancer characteristics. Titanium(IV) compounds, Budotitane [*cis*-diethoxy-bis(1-phenylbutane-1,3-dionato)titanium(IV)] [206] and titanocene dichloride (Cp_2TiCl_2), exhibit important potency against solid tumours obtained for pre-clinical trials [207–212]. The mechanism of action of titanium(IV) compounds has been studied by Sadler et al. [213, 214]. Mechanistic understandings gained for many complexes analysed include probable interaction with DNA and apoptosis induction [215–217]. Finely dispersed TiO_2 was shown to reduce the HeLa cell line [218] as well as U937 cells at UV activation [219, 220]. The ROS created by photoexcited TiO_2 can possibly damage DNA.

Recent developments in Ti anti-tumour studies open numerous guidelines to produce definite, stable in the hydrolysis interaction and active anti-cancer candidates. The main directions contain the creation of non-metallocene and non-diketonato Ti coordination compounds with bis-phenolato ligands [221], carbonyl derivatives of titanocene [222] and fulvene derivatives of titanocene [223]. Alternative current improvement in Ti-anti-neoplastic drug study is the compound Titanocene Y, a variation of dichloridotitanocene. This coordination compound has methoxyphenyl substituents on each cyclopentadienyl ring, which gives better stability, solubility in aqua solutions and better cytotoxic activity [224–226]. The SAR will be established for the numerous substituted alternatives of the initial complex.

The most important discovery in the field of titanium anti-cancer agents came with the introduction of titanocene [227]. Recently, there have been explorations into synthesizing titanocene derivatives to decrease its lipophilicity and hopefully lower the in vivo toxicity. Quieroz [228] and Tacke [229] have been involved in the more recent titanocene chemistry. The chemistry being explored by Quieroz involves modification of the Cp rings to form the complex known as Titanocene X. This new titanium derivative can be readily synthesized by first synthesizing the substituted fulvene ligand and reacting it with titanium dichloride. This new titanium dichloride derivative was tested in vivo against Ehrlich's ascites tumour inoculated mice. This data indicates that this new derivative is quite effective at certain doses in vivo. The work done by Tacke et al. involves modification of the Cp rings with dimethylamino and heteroaryl groups to try to improve the efficacy of this class of drugs. Modification of the Cp rings was explored in hopes of lowering the in vivo toxicity of these types of metallodrugs. The anti-cancer potency of these compounds was experienced against the human carcinoma cells LLC-PK to make a direct comparison to the activity of cisplatin and Titanocene X. The data presented indicates that, at least in vitro, these new titanocene derivatives are extremely active against carcinomas. Interestingly, all of these compounds are more active than Titanocene X which has already shown excellent efficacy in mouse in vivo model.

The mechanism of cancer activity for titanium is not completely understood; however, there have been some studies done to shed some light on this subject [230–232].

It is believed that titanium complexes bind to nucleic acids and therefore suppress DNA or RNA synthesis. Other groups have shown titanium to inhibit protein kinase C which is an enzyme that regulates cell proliferation. Still others have demonstrated the ability of titanium to inhibit the enzyme topoisomerase II which performs an important role in the synthesis of DNA. From all of these studies, a precise model of action was not determined, but may be clarified in the future.

Numerous novel amino-functionalized aqua-soluble titanocenes were prepared, which displayed substantial cytotoxic activity against different human tumour cells together with cisplatin-resistant cell lines like A2780-*cis* [233]. Other types of functionalized titanocenes consist of dichloride derivative compounds holding methoxy groups close to the cyclopentadienyl groups which showed potent cytotoxic behaviour against kidney carcinoma cell line (LLC-PK) and ovarian carcinoma cell line (A2780/cp70) [234, 235].

Some coordination complexes of titanium(III) were also investigated for their cytotoxic and anti-microbial activities. Ti(III) complexes were found to show moderate activity when experienced at high concentration on bacteria and fungi [236, 237]. Among the appealing properties of titanium complexes is that they do not demonstrate usual adverse effects of the known cytostatic and cytotoxic agents which make the synthesis of novel titanium compounds motivating for the possible use in the combined therapy.

10.15 Zinc

Zinc is thought to be comparatively non-toxic, particularly if administrated orally. Zinc and its inorganic salts do not cause problems to the human skin. Cases of intoxication have been caused by the ingestion of zinc inorganic salts (chloride or sulphide), but these are infrequent. Excessive inhalation of zinc compounds is frequently associated with copper shortage and can cause fever, cough and vomiting [4, 5].

Rather than being a toxic metal ion, zinc is an important trace element for all organisms including plants and animals. Among trace elements, zinc possesses a second position after that of iron in biological importance. Though the normal adult body contains only 2–3 g of zinc, the element possesses some very significant functions. Zinc participates in many reactions in the body, which help the bodies construct and maintain DNA and protein metabolism. A lack of zinc in the world is the most typical soil deficiency.

Over 300 Zn enzymes have been detected and their roles have been clarified. They accomplish every feasible type of enzyme functions. One of the most important of these functions is the participation in hydrolysis reactions; the Zn-containing hydrolases catalyse the hydrolytic reactions of P-O-P, P-O-C and C-O-C chemical bonds. That is why zinc is the most critical element in the diet. Zinc is commonly considered as a very significant element especially for the immune system, vital for the use of hormones in the body.

Zinc is needed for the cell growth, cell division and cell function. Whereas several other metals are well-known cancer-causing agents, zinc is not generally considered to be a contributing agent for cancer development. Dislocation of zinc from Zn-binding structures may be a mechanism for carcinogenicity of Cd, Co, Ni and As. The zinc ion is totally resistant to oxo-redox changes at the values of physiological potentials; therefore, Zn cannot serve a redox function, but nevertheless zinc ion is very useful. Zinc acts as a Lewis acid in enzymes. Zinc prefers tetrahedral geometries in most of the zinc enzymes. Moreover, Zn(II) undertakes very fast ligand exchange, enabling its role in enzyme functions. Systemic zinc poisonousness is not a main health problem, but due to its essentialness, an absence of this metal leads to far more severe and prevalent problems. Along with the common symptoms of zinc deficiency, the zinc shortage has been shown to induce oxidative stress, disturbance of growth and influence on the immune system [5–7].

10.16 Organozinc compounds

Organozinc reagents are a significant class of organometallic compounds with a wide variety of applications [238]. They contain carbon to zinc chemical bonds. The preparation of zinc organometallic compounds is represented by the below equation. In this interaction, the Grignard-type CH_3ZnI is an intermediate compound:

$$2CH_3I + 2Zn \rightarrow CH_3ZnCH_3 + ZnI_2$$

Zinc forms many Grignard-type compounds, such as RZnX (X = Cl, Br, I; R = ethyl, butyl, etc.). Organozinc compounds were between the first organometallic compounds prepared. They are less reactive than many other similar organometallics. Zinc organometallic compounds are similar to their corresponding magnesium compounds. Zinc organometallic compounds are highly reactive compounds and should be accorded the same carefulness to toxicity. The burning of very flammable organo-Zn substances, for instance dimethyl and diethyl derivatives, gives fine particles of ZnO fumes:

$$2(CH_3)_2Zn + 8O_2 \rightarrow 2ZnO + 4CO_2 + 6H_2O$$

Though zinc oxide is useful, inhalation of ZnO fume particles causes fever and high temperature [1–7]. Diphenylzinc illustrates the poisonousness that is obtainable from the organic part upon decomposition. At some conditions, $(C_6H_5)_2Zn$ produces poisonous C_6H_5OH and Zn species:

$$(C_6H_5)_2Zn + \{O_2\}/H_2O \rightarrow C_6H_5OH + \text{Zinc species}$$

Numerous Zn substances with organic components (organic acids salts, zinc coordination compounds) perform a significant role in various biological systems and some of them have beneficial uses including anti-fungal zinc compounds, soaps – zinc stearate and palmitate, antibacterial Zn bacitracin, low-toxic wood preservative Zn naphthenate and Zn phenolsulphonate with insecticidal possessions. Currently, new functionalized phthalocyanine–Zn(II) complexes have been estimated for their photodynamic potentials showing mediated photo-cytotoxic activity to J774 murine macrophage cells [239]. In many of the studied Zn(II) complexes, zinc is found to be bioactive interacting with target proteins. The design and screening of zinc coordination compounds with advanced activity is not efficient without consideration of the long-term toxicity, side effects and pharmacokinetic properties.

10.17 Cadmium

The toxicological effects of the group IIB metals rise in the order Zn, Cd, Hg. Compared with those of zinc, the toxic risks of Cd are relatively high. Cadmium is a poisonous metal which is present in food and is generally ingested at maximum levels. This element and its compounds are highly toxic, and contact with this metal causes cancer. It has solubility in organic acids occurred in food and produces salts that can be transformed to $CdCl_2$ through the gastric juices. Cadmium is a respiratory toxic element. Acute exposure to Cd fumes may cause flu-like symptoms. Chronic exposure can result in kidney, bone and lung diseases [4, 5]. The kidney is most disposed to cadmium, around 200 ppm can cause serious damage. Smokers also absorb substantial amounts of cadmium. Contact to cadmium from manufacturing sources especially the Ni–Cd battery is becoming a major problem. There are varied thoughts on Cd – while it is certainly believed not to be crucial for living physiological processes, some trust Cd has a necessary role in metabolism and other biological and pharmacological processes. Its position as a vital trace element still remains uncertain, and its necessity and use are not presently understood.

10.18 The toxicity of cadmium

Cadmium poisoning affects many organs. Along with metals like Hg and Pb, this metal is one of the very hazardous metal toxicants. Cadmium belongs to the same group IIB in the periodic table below zinc and conduct yourself like zinc. This is a reason for cadmium's poisonousness; Zn is a crucial trace element, consequently Cd substitution for zinc could disturb many biological processes.

Cadmium is known with its eight stable isotopes between ^{106}Cd and ^{116}Cd. The most common of them are ^{112}Cd and ^{114}Cd. Cadmium produces a variation of complex organic complexes with amine, sulphur and chloro substituents. In all of its compounds, this element occurs in the form of Cd^{2+} ion. Cadmium ions produce soluble inorganic salts such as carbonates, arsenates, phosphates and ferrocyanide salts. Associated with the zinc production, it can be obtained in variety of industrial forms.

Cadmium disturbs cell production, differentiation as well as apoptosis. These processes are connected with the ROS generation resulting in oxidative stress and the apoptosis induction. Cadmium constrains the potency of enzymes with antioxidant properties, for instance catalase, MnSOD and Cu/Zn-dismutase. Cadmium can also modulate the cellular level of Ca^{2+}. This element has poor absorption by the gastrointestinal tract. It is transported to red blood cells and is excreted by urine and faeces.

Different forms of cadmium compounds have various clinical demonstrations and poisonous effects. Acute pulmonary symptoms of Cd contact can be caused by the inhalation of CdO, which results in pneumonitis and other pulmonary diseases. Chronic exposure causes emphysema. Cadmium mainly gathers in the kidney, liver, bone, etc. The kidney is the most sensitive organ to chronic Cd intoxication. Chronic poisonous effects of cadmium contact include skeletal demineralization, as well as hypertension and adverse cardiovascular effects. Itai-itai disease is the most serious form of chronic cadmium toxicity [5–7]. Cadmium compounds were categorized as carcinogenic in humans, including lung, prostatic and renal cancers. This metal is highly cumulative and its half-life is around 20–30 years.

Cadmium in the body affects several enzymes [1–7]. The toxicity of Cd^{2+} is connected with its high thiophilicity, permitting Cd^{2+} ions to replace Zn^{2+} in the respective enzymes. Cadmium(II) has a larger ionic radius than Zn(II) and Lewis-acid properties, and that is why a substitution of Zn^{2+} for Cd^{2+} generally results in an inactivation of the respective enzyme. Along with this, because of the comparable ionic radii of Cd(II) and Ca(II), cadmium ions also act as Ca^{2+} antagonists. Cadmium ions can be constructed into Ca sites of bone hydroxyapatite, causing osteoporosis. Additionally, cadmium ions have a high attraction to the phospholipids in cell membranes. Through the coordination interactions, this disables the functions of membranes.

The removal of heavy metals includes the addition of some non-toxic chemicals to decrease the solubility of the respective metal. Elimination of cadmium or detoxification can be achieved by coordination to thioneines, by glutathione γ-Glu-Cys-Gly or by phytochelatines $\{(\gamma\text{-Glu})\text{-Cys}\}_n$-Gly (where $n = 2–11$). Cadmium intoxications need refinement, chemical decontamination and the usage of nanoparticles and chelating agents [240–243].

10.19 Cadmium organometallic compounds

Organocadmium compounds contain carbon–cadmium bond. The corresponding chemistry of cadmium, zinc and mercury is very similar. In the absence of water, cadmium halides interact with organolithium compounds:

$$CdBr_2 + 2Li^+C_6H_5^- \rightarrow 2LiBr + (C_6H_5)_2Cd$$

Chemical interactions of organocadmium compounds with imine, nitrile and carbonyl compounds are not as typical as for the equivalent organometal compounds of zinc.

The oily liquid $(CH_3)_2Cd$ is poisonous to the respiratory system at inhalation. Diethylcadmium ($(C_2H_5)_2Cd$) is like oil and interacts actively with O_2 in air. $(C_3H_7)_2Cd$ is a typical oil and interacts with H_2O. The dialkyl analogues are not stable above 150 °C and undergo decomposition.

A major weakness of the use of organocadmium reagents is the poisonousness and environmental problems associated with usage of cadmium, and this has limited the recent use of organocadmium reagents [5–7]. The cadmium organometallic compounds produce vapours, containing Cd and CdO, which are toxic at inhalation [243, 244].

10.20 Mercury

In contrast with cadmium, contamination by Hg, although theoretically a serious problem, was not found as a comprehensive problem [1–7]. Mercury sources are combustion of waste from Hg batteries, the chlor-alkali electrolysis by the amalgam process and pesticides. Mercury and its inorganic compounds are extremely poisonous, and its organic compounds in addition are teratogenic. They are accountable in the atmosphere, aqua-sphere and sidero-sphere. The species can be abiotic and biotic.

Mercury is directly below cadmium in group IIB, but has significantly different chemical properties than its analogues Cd and Zn [5–7]. Elemental mercury is the only liquid metal at normal conditions. Having a high vapour pressure, this element is very dangerous, especially to the central nervous system. The metal has slight solubility in water because of the weak metal bonding.

Mercury occurs in various compounds: metallic Hg, inorganic compounds and organometal compounds (methylmercury). These mercury compounds vary in the degree of poisonousness and in the consequences on many systems. Mercury occurs naturally mostly in metallic form and as cinnabar (HgS). It is mainly produced from volcanic activity. Inorganic mercury compounds are typically less toxic because of their low solubility. More hazardous are their organometal compounds. They can be easily absorbed and kept by the organism more strongly than the inorganic analogues. The methylation of Hg in biological environment is performed by methylcobalamin, the important vitamin B12. Another major hazard is the organomercury fungicides.

Human activity remains the key source of Hg pollution, for instance the production of thermometers and barometers. Currently, mercury is used in chlor-alkali electrolysis, in the compositions of dental fillings. As mentioned above, mercury can be converted to methylmercury by some bacteria. Methylmercury then bioaccumulates in organisms.

Along with the usages of the metal, the compounds of mercury have numerous applications. HgO is the main raw material for the production of additional Hg compounds. Mercury(II) acetate, $Hg(C_2H_3O_2)_2$, has a good solubility in many organic solvents. The water-soluble Hg(II) chloride, $HgCl_2$, is very poisonous in comparison with the other inorganic compounds. In contrast with the group analogues, Hg produces also monovalent substances in the form of binuclear Hg_2^{2+} ion. The most important mercury(I) compound is calomel, Hg_2Cl_2, used as a component of electrodes. Hg(I) compounds are commonly less poisonous than Hg(II) compounds due to their lesser solubility. There are many toxic organomercury compounds. It has to be mentioned that Hg compounds are used as disinfectants ($HgCl_2$). Inorganic mercury is a component of some skin-treating products. Many mercury compounds of the aryl and alkyl groups are used as fungicides. The highly toxic mercury compound methylmercury is a potent neurotoxin [4, 5, 245]. Thiomersal (ethylmercury) has preservative properties and its main usage is in the pharmaceutical industry. It does not accumulate by the organism and is not a health hazard. Mercury and its compounds can be touched safely, but strict precautions must be taken at inhalation and ingestion.

Metallic mercury in its vapour state can be easily absorbed by the pulmonary system. Inorganic compounds are usually absorbed by the intestinal route or by the skin [5–7]. Mercury exposure can lead to developmental problems in the brain, which can also affect physical functions. Though metallic mercury is quickly oxidized to Hg(II) in the red blood cells, a huge part of the metal reaches the brain before oxidation which is connected with the lipid solubility of metal Hg(0). Inorganic Hg(II) ions and the respective compounds have a tendency to accumulate in the kidney. Like its analogue Cd, Hg(II) has a strong attraction to sulphhydryl groups of many proteins, enzymes, haemoglobin and serum albumin. Sulphhydryl groups are abundant as active functional groups of many enzymes, and that is why it is problematic to establish precisely which of them are affected by Hg in living systems. The consequence of mercury vapour on the central nervous system is largely psychopathological (tremor, emotional instability and depression). These indications are the consequence of impairment to the blood–brain barrier, which controls the transmission of metabolites [1–7].

10.21 Organomercury compounds

The first synthetic organomercury compound has been made by E. Frankland in 1853 by the reaction at photochemical condition:

$$2Hg + 2CH_3I + h\nu(\text{sunlight}) \rightarrow (CH_3)_2Hg + HgI_2$$

Many synthetic paths are possible for the synthesis of various Hg organometallic compounds. Later, many of these medicinal compounds were obtained and used. These organomercury compounds have been widely used as pesticidal fungicides [246].

Most literature data on organomercury compounds are connected with methylmercury derivatives, for example monomethyl salts (CH_3Hg^+) and dimethyl salt $(CH_3)_2Hg$, which have good solubility and are instable. These kinds of compounds are obtained from inorganic Hg by means of anaerobic bacteria with the help of methylcobalamin:

$$HgCl_2(s) + (\text{Methylcobalamin}) \rightarrow CH_3Hg^+ (aq) + 2Cl^-$$

The interaction is favoured in acidic solutions. In neutral or slightly alkaline water solution, dimethylmercury production is favoured. Dimethylmercury is volatile and may easily reach the atmosphere. Additionally, these compounds have lipid solubility, favourable for bioaccumulation through the pulmonary system, as well as by the skin and gastrointestinal tract.

Regarding the biodistribution, the methylmercury compounds are similar to metallic Hg, because the deposited amounts of them can be converted to metallic mercury. Corresponding to elemental Hg, they cross the blood–brain barrier, thus disturbing the nervous system. Methylmercury interacts with sulphhydryl groups and produces similar signs and symptoms as ethyl mercury [5–7].

Chapter 11
Transition metals in chemotherapy

The affiliation of metals to cancer treatment combines the expertise of many specialists in the field of bioinorganic chemistry, pathology, pharmacology and oncology [1–7]. Therapeutic activity of metal coordination compounds in cancer remedy is interesting mostly because of the unique features of metals, including their activity in oxo-reduction processes, flexible coordination structures and chemical reactivity with bio-substrate. Redox-active metals commonly produce ROS, and they are used to induce DNA cleavage [247, 248]. Ligand exchange and variation of the initial chemical constructions led to the preparation of many metal-based complexes with improved cytotoxic activity and different pharmacokinetic profiles. Several metals of interest (e.g. gold, platinum, ruthenium, rhodium and silver) and their metal-based compounds which have been examined for effective treatment of cancer or for active reduction of the malignancies are described in this chapter.

11.1 Gold

The usage of gold-based compounds in medicine (chrysotherapy or aurotherapy) started in 1935. Some gold(III) and gold(I) compounds have been considered for their medicinal effectiveness in numerous diseases, particularly rheumatic arthritis. The cytotoxic activity of Au complexes has attracted consideration as an alternative to cisplatin. The electronic configuration of Au(III) is d^8, equivalent to Pt(II) and helps the square-planar geometry typical for Pt(II) complexes. Many structures of gold(III) complexes have been obtained, and their anti-neoplastic activity has been evaluated against different cancer cells. The physiological and anti-proliferative potencies of gold(III) complexes do not stand up from the attraction to DNA, and this is rather different from representative Pt(II) complexes [249]. In addition, the anti-tumour Au(III) complexes hardly disturb the cell cycle [250]. Current investigational suggestions specify a direct interfering with mitochondrial functions [251, 252].

Most of the described gold(III) complexes have a strong effect on cell lines resistant to cisplatin. The preferred donor centres and the ligands are Cl, Br, S and P. Mononuclear anti-neoplastic gold(III) complexes can be generally classified in the next group: gold(III) polyamines [253], gold(III) polypyridines [254, 255], gold(III) porphyrins [256], gold(III) dithiocarbamate [257–259] and organogold [260] compounds. None of these complexes has made it to clinical trials [261] despite their good redox activity, in vitro cytotoxic effects and different from Pt(II) mechanism of action.

The use of gold in the field of treating cancer dates back to the late 1980s with the work reported by Mirabelli [262]. There have been some reports on the activity of Au(I) complexes [263, 264], and the most recent work in this area has been explored

https://doi.org/10.1515/9781501516115-015

by Filipovska [265] and Reedijk [266]. The work studied by Filipovska involves the use of Au(I) complexes of phosphines, which have been previously shown to have promising anti-cancer effects [267]. The previous work employed the use of diphenylphosphine ligands to complex the gold atom, whereas in the newer work dipyridylphosphines have been used as the ligands of choice. The change in ligand was a result of the in vivo hepatotoxicity that was found to occur with diphenylphosphine complexes in clinical development. The gold complexes were examined against the human breast cancer cells MDA-MB-468 and the normalized breast cell line HMEC (normal human mammary epithelial cells) to determine their anti-cancer activity and selectivity between normal and cancerous cells.

Though gold has been used clinically for 80 years, its mode of action is not fully elucidated [2]. However, the mode of action is believed to be connected to the coordination at the gold centre. It is believed that tetrahedral gold complexes attack the mitochondria of the cell. Studies have been done in rat hepatocytes with tetrahedral gold phosphines. The mitochondria were isolated, and it was determined that much of the drug was uptaken due to its cationic and lipophilic nature. Once in the mitochondria it is believed that gold attacks the mitochondrial DNA producing cross-links and strand breaks, along with inhibition of protein synthesis, efflux of calcium and mitochondrial swelling. These effects eventually lead to the shutdown of the mitochondria and cell death [267, 268].

It is well known that Au compounds are predominantly effective in the rheumatoid arthritis treatment and in the anti-inflammation; however the mode of action is still discussed. Some Au(I) complexes were studied for their anti-inflammatory effects in the treatment of rheumatoid arthritis [269]. It was observed that people experiencing chrysotherapy have low risk of malignancy [270]. Several well-recognized drugs like cyclophosphamide, 6-mercaptopurine and methotrexate show both anti-neoplastic and anti-inflammatory activities [271–273]. One of the most common compounds used in the rheumatoid arthritis treatment is auranofin, $(C_{14}H_{19}O_9)$-S-Au = $P(C_2H_5)_3$. Auranofin displays activity against Hela cell line [274] and against P388 leukaemia cell line in vivo [275]. A number of phosphane-gold(I) complexes were potent when tested in vitro against P388 and B16 cell lines and in vivo against P388 [276]. The connection of organophosphanegold(I) with bioactive thiols perhaps displays double action of both parts and the cytotoxic in vitro and in vivo profile is improved [277, 278]. The bidentate phosphanes form a coordination compound with gold(I) which have effective cytotoxic activity as observed by means of in vitro and in vivo assays. The clinical trials have been stopped as a result of acute toxic effects to many organs like lungs, liver and heart [279–281].

Numerous binuclear gold(I) complex compounds of bidentate nitrogen–heterocyclic carbene ligands were estimated [282]. The effects of several gold(I) and gold(III) complex compounds on mitochondrial thioredoxin reductase were studied. Both Au(I) and Au(III) complexes were revealed to be very effective [283]. Novel gold(I) and gold(III) complex compounds of chloroquine were in vitro estimated against *Plasmodium falciparum* and have showed good anti-malarial features against chloroquine-sensitive and chloroquine-resistant strains [284, 285].

11.2 Rhodium

Rhodium was discovered to have anti-cancer properties in the early 1970s when Bear and co-workers were exploring the activity of rhodium(II) carboxylates against various tumour cell lines [286]. These structures are generally tetrabridged systems with short Rh–Rh bonds and contain various axial ligands. Pruchnik [287] and de Souza [288] further explored rhodium(II) carboxylate derivatives for their anti-cancer activity. The work explored by Pruchnik involves not only rhodium(II) carboxylates but also rhodium(II) complexes with the nitrogen containing ligands 2,2′-bipyridine and 1,10-phenanthroline. A major drawback of these types of rhodium complexes is their limited water solubility. Therefore, de Souza and co-workers explored the effect of axial ligand coordination on the solubility of these rhodium complexes. The rhodium complexes showed anti-neoplastic activity against oral carcinoma KB cell line. These rhodium complex compounds were obtained by interaction of $Na_4[Rh_2(CO_3)_4]$ with acetic acid, with formic acid and 2,2′-bipyridine, and with formic acid and 1,10-phenanthroline [289, 290].

De Souza and co-workers explored similar structures to Pruchnik; however, they were interested in improving the water solubility of the dirhodium carboxylates by varying the axial ligands bound to the rhodium centre. To enhance water solubility, the axially coordinated water ligands were replaced with isonicotinic acid [291]. These new dirhodium carboxylates were synthesized by slowly adding two equivalents of isonicotinic acid in methanol to a stirred solution of the previously synthesized dirhodium carboxylates in methanol [292]. Rh(II) complex compounds were experienced for their anti-neoplastic activity against K562 cells, a human leukaemia cell line and for their acute in vivo toxicity in male Balbc mice. For the acute toxicity study, the mice were broken into four groups of eight per complex. Each group was given an IP injection of a different dose to determine the LD_{10} of each complex. Because there are reports that dirhodium compounds inhibit protein synthesis and DNA replication comparable to cisplatin [293–295], the mechanism by which these dirhodium carboxylates cause cell death has been extensively studied by Chifotides and Dunbar [296, 297].

Rhodium(III) complexes have also been studied, though to a much lesser extent than rhodium(II), for their anti-cancer activity [298–301]. The most promising rhodium (III) anti-cancer agents have been explored by Pruchnik where the ligands pyrazole, imidazole, triazine and terpyridine have been used to complex rhodium(III) [302]. The rhodium(III) complexes were obtained by adding an ethanolic solution of the respective ligand to $RhCl_3$ in ethanol. All of the products precipitated as the yellow solids.

These rhodium(III) complex compounds were experienced for in vitro anti-neoplastic activity against bladder cancer cells HCV29T and compared to the activity of cisplatin.

Rh radiopharmaceuticals have also been reported. Novel ligand groups for stabilization of ^{105}Rh radiopharmaceuticals involved *N*-heterocyclic carbenes obtained by a direct interaction of the carbene with $RhCl_3{\cdot}xH_2O$ [303]. Alternative study demonstrated the anti-cancer activity of new Rh(III) complex [*cis*-dichloro(dipyrido {3,2a-2′,3′c}phenazine)(1,10-phenanthroline)rhodium(III) chloride] [304].

11.3 Ruthenium

Before the discovery of cisplatin, the best-known chemotherapy drug, some Ru-based complexes have been tested for their anti-tumour potentials [305]. Ruthenium is located in group VIIIB in the second period of the periodic table, below Fe. Ruthenium has a good potential to form coordination compounds because of its wide range of easily obtainable stable oxidation states between −2 and +8. Ruthenium can form complexes with octahedral geometry that provide prospect for using more organic ligands in comparison to Pt(II) compounds that only form complexes with square planar geometry. In water solution, its most stable oxidation states are +2 and +3. These compounds possess relatively high kinetic stability which make ruthenium compounds appropriate for medical applications. First of all, the rate of ligand exchanging reactions is comparatively slow and similar to that of Pt(II) and Pt(IV) [306]. Its accessible oxidation states (+2, + 3 and + 6) are stable under physiological conditions. Additionally, ruthenium mimics iron in binding to certain life molecules [307]. The most important is the octahedral geometry of ruthenium, related to square-planar geometry of Pt, and the possibility of a dissimilar mechanism of action with a diverse capacity.

Currently, the therapeutic chemistry of ruthenium has extended to a broader range and Ru compounds have been examined as immunosuppressants, antibiotics, NO scavengers and anti-tumour and anti-metastatic drugs [308–310]. The earliest reports of ruthenium complexes possessing anti-cancer activity came in the 1980s when Clarke discovered the activity of *fac*-[Ru(III)$Cl_3(NH_3)$]; however, this drug has poor water solubility [311]. To date, there have been a variety of ruthenium complexes studied; however, only a few complexes have had activity comparable to cisplatin. The most promising class of ruthenium compounds is represented by Ru(III) salts with imidazole and indazole ligands connected to the metal centre by their respective free nitrogen atoms. Keppler was the first to report on these types of complexes [312]. Impressively, they proved to be quite efficacious against colorectal tumour cells and many resistant to cisplatin tumours with small toxic adverse effects [313]. The most impressive characteristic of the complexes is that they exceeded the activity of 5-fluoruracil, the standard treatment for colorectal cancer today. These results have led to the complex being explored in clinical trials. So far, data from phase I and II have shown extremely promising results against colon, liver, head, neck, etc. cancers with active doses being well tolerated.

Mononuclear anionic ruthenium compounds are labelled as NAMI and NAMI-A, where NAMI is the abbreviation for *New Antitumour Metastasis Inhibitor*. The most important of them are *trans*-Na[$RuCl_4$(dmSo)(Him)] (Him = imidazole) and *trans*-(H_2Im)[$RuCl_4$(dmSo)(Him)] [314, 315]. NAMI showed activity meaning better than cisplatin on Lewis lung carcinoma, MCa mammary carcinoma or B16 melanoma, as well as a drastic decrease of the metastasis's formation. The low stability upon storing of NAMI led to the synthesis of NAMI-A, passed in clinical trials. NAMI and NAMIA significantly increased the width of the connection materials of the tumour capsule and round tumour blood vessels, which possibly delays metastasis expansion and blood flow to the tumour [316–319]. NAMI-A has shown very impressive anti-cancer and anti-metastatic properties [320]. This ruthenium complex was first synthesized because ruthenium coordination compounds with dimethylsulphoxide have been discovered to have anti-neoplastic and anti-metastatic activities. The remarkable preclinical activity of NAMI-A has led to its use in clinical trials and this potential drug has completed phase I as an anti-metastatic agent.

More recently, there has been work done exploring the activity of ruthenium(II) complexes. Sadler [321] has explored ruthenium arene complexes and Reedijk [322] has looked into ruthenium complexes of 2-phenylpyridine. Both of these categories of complexes seem to be active against human ovarian carcinomas.

The work explored by Sadler involves Ru(II) complex compounds with the structure [(η^6-arene)-Ru(A)(B)(C)], where the arene ligand is benzene/substituted benzene, and A, B and C are monodentate halides, acetonitrile and nicotinamide, or A–B is a bidentate ligand such as ethylenediamine or *N*-ethylethylenediamine. The arene ligands are believed to help in the stabilization and most of these complexes are ionic leading to enhanced water solubility. The metal complex compounds were obtained by adding the respective [(η^6-arene)-Ru(A)$]_2$, where A is a halide, to either acetonitrile, methanol or benzene, and allowing them to stir with NH_4PF_6, or adding in a diamine or isonicotinamide and letting them stir with NH_4PF_6. The anti-neoplastic activity of the complexes was tested against the human ovarian cancer cells A2780. The IC_{50} values were calculated after exposing the cells for 24 h with each Ru(II) complex. None of the complexes is as active as cisplatin; however, some of them are as active as carboplatin. The mechanism by which these Ru(II) complexes kill cells has not fully been determined, but preliminary reports by Sadler indicate that they selectively bind to the N7 site of guanine.

The Reedijk group has been working with the water-soluble bis(2-phenylazopyridine)Ru(II) complex compounds to determine their level of anti-cancer activity. These metal complexes can be readily synthesized by adding the ligand 1,1-cyclobutanedicarboxylic acid, oxalic acid or malonic acid to a stirring solution of α-[Ru-(azpy)$_2$(NO_3)$_2$], where azpy is 2-phenylazopyridine, in acetone at 40 °C for 4 days. The anti-neoplastic activity of these complex compounds was also established against the ovarian cancer cells A2780 along with the cisplatin-resistant type of these cells A2780*cis*R. The cells were incubated for 72 h with the respective metal complexes, and the IC_{50} values were then determined. All the complexes were more active than carboplatin and on the order of cisplatin against the normal cancer cells. Impressively, they were all more effective than both cisplatin and carboplatin against the resistant cell lines. The stability of these compounds was examined by ^{1}H-NMR spectroscopy under mimicked physiological conditions where $CDCl_3$ was modified to be at pH 7.4 by the addition of phosphate buffer and contain NaCl concentration of 0.1 M. These are the conditions of physiological saline solution and are the best mimic for in vivo bloodstream conditions. The spectra were monitored for 1 month to see what effects were caused. Metal complexes showed no change in their respective spectra, meaning they would probably be stable in vivo.

The mechanism by which ruthenium is active against cancerous cells is believed to occur through one of three different pathways. These three mechanisms include the construction of Ru-DNA inter-strand cross-links, Ru mimicking Fe and interaction with transferrin, and lastly many rapidly dividing cells are hypoxic and have lowered pH which are favourable conditions for the Ru interaction with intracellular proteins [2]. Reedijk and co-workers have explored how ruthenium bound to guanine [323]. They used the ruthenium compound *mer*-[Ru(terpy)Cl_3] and reacted it with a variety of guanine derivatives and discovered that these types of ruthenium compounds formed interstrand DNA cross-links with the N7 position of the guanine ring. This is interesting because it is opposite of the intrastand cross-links formed by cisplatin. Clarke and Keppler explored the ability of ruthenium to mimic iron [324]. Because rapidly dividing cells require more iron, they have high number of transferrin receptors on the surface. Work done in this area has shown that ruthenium binds to transferrin and therefore it is believed that transferrin carries ruthenium into the cell which may help explain its low toxicity in vivo. Clarke and Keppler have also explored why the hypoxia and lowered pH allow for Ru(III) complexes to be active in cells. They have shown that this is due to the ability of Ru(III) to readily undergo reduction to Ru(II) in these types of environments. Therefore, Ru (III) drugs are believed to be prodrugs and it is actually Ru(II) that is the active species. This mechanism is known as "activation by reduction."

Ru(III) compounds with heterocyclic monodentate ligands showed promising anti-cancer properties. These are *trans*-(H_2Im)[$RuCl_4$(Him)$_2$] (Ru-im or ICR) and *trans*-(H_2Ind)[$RuCl_4$(Hind)$_2$] (also called Ru-ind or KP1019). These compounds exhibited marked potential against P388 cell lines, while Ru-im showed improved action against Walker carcinoma cell lines and Stolkholm ascitic tumour [325, 326]. In an induced colorectal tumour rat model (which is insensitive to cisplatin), the compounds Ru-im and Ru-ind exhibited proficient results. Ru-ind was found less toxic and can be applied at higher dose. The presented anionic coordination complexes are typical pro-drugs which hydrolyse quickly in vivo to produce comparatively stable and neutral *trans*-[$RuCl_3$(H_2O)(L)$_2$] compounds.

Other novel ruthenium complexes that were obtained and showed cytotoxic properties on different tumour cell lines include *cis*-$[RuCl_2(dmSo)_3(dmsO)]$ as anti-tumour and anti-metastatic agents [327]. It decreases significantly primary tumour growth in the tested tumours (Lewis lung carcinoma, B16 melanoma and MCa mammary carcinoma). The respective *trans*-$[RuCl_2(dmSo)_4]$ showed also proficient anti-tumour and anti-metastatic properties at lesser doses related to its *cis*-isomer. The *cis*-isomer in water directly undertakes loss of the dms*O* ligand, while the *trans*-isomer produces a *cis*-di-aqua complex, losing two dm*S*o ligands. Both hydrolysed isomers then undertake slow dissociation to chloride, thus producing cationic species. The *trans*-isomer with three active species is more potent compared to the *cis*-isomer.

Among the Ru-ammine compounds, *cis*-$[RuCl_2(NH_3)_4]Cl$ and *fac*-$[RuCl_3(NH_3)_3]$ display an encouraging response against P388 leukaemia cell line. $[RuCl(NH_3)_5]^{1+/2+}$ and $[Ru(H_2O)(NH_3)_5]^{2+/3+}$ bind favourably to guanine and adenine. The pyridine analogue *cis*-$[Ru(H_2O)(py)(NH_3)_4]^{2+}$ binds completely to the guanine residue [328].

Newly synthesized organo-ruthenium compounds have been widely investigated for their anti-cancer activity. The organometal compounds, having the formula, $[Ru(II)X(\eta^6\text{-arene})(en)]^+$ (X – halide, arene, e.g. benzene/substituted benzene, en – ethylenediamine), of this class have been synthesized and were found to have promising in vitro IC_{50} values in the A2780 cells [329–332]. The most potent compound of this group, $[Ru(\eta^6\text{-tha})(en)Cl]PF_6$ (tha = tetrahydroanthracene), has equipotent activity as the drug cisplatin [333]. It was observed that cytotoxic activity raised by growing the hydrophobicity of the respective arene. These complexes were active against some resistant to cisplatin tumour cell lines. Substituting ethylenediamine with larger N-donor ligands, for instance bipyridine or *N,N,N′,N′*-tetramethylethylenediamine, decreases the potency [334].

Ru(II)–arene complex compounds with amphiphilic ligand 1,3,5-triaza-7-phosphaadamantane (PTA), namely $Ru(\eta^6\text{-toluene})\text{-}(PTA)Cl_2$, RAPTA-T, $Ru(\eta^6\text{-}p\text{-cymene})(PTA)Cl_2$ and RAPTA-C, showed small toxic effects in vivo, but as a whole, they displayed related anti-neoplastic activity. The complex, $[Ru(\eta^6\text{-p-cymene})(PTA)Cl_2]$, soluble in water, exhibited DNA binding, depending on pH [335]. Comparable to NAMI-A, the compounds were not active against primary tumours, but were recognized to be active against metastases in vivo. Given the structural modifications between RAPTA and NAMI-A complexes, they are expected to interact in a different way with proteins, to select dissimilar biomolecular targets or to have different mechanism of action. A series of water-soluble dinuclear Ru(III) complexes, such as $Na_2\{(trans\text{-}RuCl_4(dmSo))_2(\mu\text{-}L)\}$, (L is an aromatic N-donor ligand), have been obtained and their anti-cancer potency against animal tumour models has been investigated. In vivo, these compounds demonstrated activity comparable to that of NAMI-A against MCa mammary carcinoma [5–7].

11.4 Rhenium

There have been a variety of reports on the activity of rhenium in the field of cancer therapy [336–340]. More recently, work in this area has been explored by Moreno-Carretero [341] and Wong [342], where both groups use Re(I) as the metal of choice. The chemistry explored by Moreno-Carretero involves complexes of rhenium with 6-amino-5-nitrosouracil ligand derivatives. The rhenium complexes can be readily synthesized by the equimolar reaction of the respective ligand and $ReCl(CO)_5$ in CH_3CN at reflux for 3 h. These rhenium complex compounds were experienced against the breast cancer cells MCF7 and EVSA-T, the neuroblastoma cell line NB69, the glioma cell line H4 and the bladder carcinoma cells ECV. The IC_{50} values of these complexes were all around 10 μM for all cell lines tested indicating that there is little selectivity for the Re(I) 6-amino-5-nitrosouracil complexes. The rhenium complexes explored by Wong involve complexation to the ligand 2-amino-4-phenylamino-6-(2-pyridyl)-1,3,5-triazine known as appt. The Re(I) complex of this ligand was synthesized by the reaction of $ReCl(CO)_5$ with appt in equimolar amounts in methanol for 24 h forming the pure yellow solid. The anti-cancer properties of this complex were explored in vitro against the epidermal carcinoma cells KB-3-1 and its multi-drug resistance analogue KB-V-1, the hepatocellular cancer cells HepG2, the cervical cancer cells HeLa and the lung fibroblast cells CCD-19Lu, compared to cisplatin. Interestingly, there is selectivity between the cancerous cell lines and the normalized lung fibroblasts, but the Re complex is not nearly as active as cisplatin.

A recently published report by Baird on the mode of action of rhenium complexes against cancer cells in vitro elucidated some interesting results [343]. They explored the ability of certain rhenium compounds to inhibit cathepsin B. Cathepsin B is a protein that is believed to be elevated in tumours and is proficient for degrading components in the extracellular matrix. In cancer cells, especially, it is believed to be involved in metastasis, angiogenesis and tumour progression [344]. The data collected by Baird indicate that rhenium complexes have the ability to actively bind the active site of cathepsin B rendering it inactive indicating that this is, at least, one way that rhenium is active against cancer.

Within the last years, novel radionuclide therapies were developed as resourceful devices for tumour and inflammatory treatment. Rhenium has two possible advantageous isotopes, ^{186}Re and ^{188}Re. Rhenium-186 and Rhenium-188 have been used in radioimmunotherapy [345–349]. Rhenium, copper and technetium complexes with different organic ligands have also been investigated as labelling agents [350–357] and have confirmed the feasibility and clinical helpfulness for a wide range of diseases.

11.5 Silver

Silver is a promising metal in cancer therapy because its toxic effects in humans are supposed to be quite low [358]. In fact, silver has been found in 29 different human tissues in trace quantities [359]. Though it is found in the body, there is no recognized physiological action for silver. There are already a variety of ways that silver is being used in medical applications, including coating of heart valves, cardiac catheters and urinary catheters to reduce or prevent infection [360, 361]. The only major known side effect of silver ingestion is a permanent skin discoloration, the so-called argyria. Argyria is not thought to be damaging to the body physically, and it takes an extreme consumption of silver to progress this condition [362].

Research efforts have been focused on developing silver(I) complexes of a variety of different ligands to combat cancer. These include silver(I) coordination complexes of phosphines, carboxylates, thio-groups and thioamides, tripodal thio-glycosides and the natural organic compound coumarin [363–370]. These silver complexes show a high level of anti-neoplastic activity against various different cell lines. The silver(I) phosphine complexes have been explored by Berners-Price and by McKeage. These silver phosphine complexes can be synthesized by the general route of adding two molar equivalents of the phosphine ligand to one equivalent of silver nitrate in either chloroform or acetone. They have been shown to be stable in light and are soluble in a variety of solvents. ^{31}P NMR spectral studies show that these silver complexes are stable in solution for at least 24 h and when nine molar equivalents of NaCl was added, the spectra were unaffected indicating the chloride does not displace the silver ion from the ligands. Stability to NaCl solution is important because there is a high concentration of chloride in the bloodstream, so potential drugs need to be stable to chloride. Other ^{31}P NMR spectral studies were done to determine the effect of glutathione (GSH) on these complexes. In the body, GSH is involved in many redox reactions including playing a major role in detoxification of harmful materials, especially the chelation and removal of heavy metals. The anti-tumour potency of these compounds was determined in vitro against the B16 melanoma cells, and in vivo against P388 leukaemia and M5076 reticulum cell sarcoma in mice. The silver complex activity was compared to the respective free phosphine ligand. The level of in vivo activity was determined by the per cent increased lifespan (ILS) of the tested animals compared to the untreated animals where the silver complexes were administered on days 1–5 following the implantation of the tumours. The ILS represents how much longer the tested animals lived compared to the untreated animals. The data indicate that the compounds that were tested in vivo were all relatively active.

Silver(I) polymeric complexes have also been studied to determine the extent of their anti-cancer properties. Zhu has explored silver(I) carboxylates and Zachariadis has explored silver(I) polymeric complexes formed with the use of conjugated heterocyclic thioamides. The polymeric structures synthesized by Zhu have the structures $[Ag(fbc)]_n$, $[Ag_2(cpd)]_n$ and $[Ag_2(idc)]_n$ (fbc = 4-fluorobenzoate, cpd = cyclopentane-1,1-dicarboxylate, idc = iminodiacetate). These silver(I) extended networks are synthesized by dissolving Ag_2O and the carboxylic acid, 1:2 for cpd and idc and 1:1 for fbc, in an ammonium solution. The solutions were allowed to sit in the air until a solid precipitate formed yielding the desired silver complexes which were confirmed by elemental analyses and IR spectroscopy. These silver polymers were tested in vitro against solid tumour cells: HeLa, HepG2, BGC, 95-D, CNE, and normalized cells: NIH 3T3 and L-02. The research explored by Zachariadis involves the self-assembly of polymeric silver compounds using the thioamide 2-mercapto-3,4,5,6-tetra-hydropyrimidine. Use of the thioamide was chosen because these types of ligands have a tendency to form bridged oligo and polynuclear compounds, a goal of this field of research. The in vitro anti-neoplastic activity was estimated against the murine leukaemia cell line L1210 and was compared to their cytotoxicity against the normal T-lymphocyte cells Molt4/C8 and CEM. The ligand itself shows no cytotoxicity to any of the cell lines tested; however, the Ag(I) complexes show good activity with some selectivity.

The anti-neoplastic activity of silver(I) complexes of coumarins has been explored by Egan [371]. Coumarins are a large class of naturally occurring compounds. Naturally, they have been shown to possess minimal anti-cancer activity; however, when complexed to metals their activity greatly increases. The silver(I) complexes of coumarins were synthesized by deprotonation of the respective coumarin derivative in an ethanolic solution of sodium hydroxide by the interaction with $AgNO_3$. The silver coordination compounds precipitated out of solution and were isolated without further purification and their structures were confirmed by IR and NMR spectroscopies. Four different coumarin derivatives were used to synthesize the silver(I) complexes 6-hydroxycoumarin-3-carboxylatosilver, 7-hydroxycoumarin-3-carboxylatosilver, 8-hydroxycoumarin-3-carboxylatosilver and coumarin-3-carboxylatosilver. The anti-cancer potency of coumarins and their silver complexes was determined in vitro against the cancerous cells A-498 (kidney adenocarcinoma) and Hep-G2 (hepatocellular carcinoma), and compared to the normal cell lines HK-2 (proximal tubular) and CHANG (hepatic) cells. The activity of the complexes was compared to silver perchlorate and cisplatin. The silver coordination complexes were as potent as cisplatin against the cancerous cells (Hep-G2 and A-498), and showed better selectivity against the normalized cell lines (CHANG and HK-2). The exact mechanism of how silver is active against the various cancer cells is not totally known; however, there have been a few reports that indicate possible answers to this question [372]. The work in this area done by Egan using their coumarin silver complexes suggest that silver kills through an apoptotic pathway. Through electrophoresis studies, they observed a DNA ladder pattern which is consistent with drugs causing apoptosis.

Another indication that silver causes apoptosis is that the cells that undergo this type of death usually have increased levels of caspases 3 and 9. Caspases are known as executioner proteins meaning they play a major role in cell death. The cells that were treated with the silver coumarins had elevated levels of both caspases. The Egan group also explored if silver causes defects in the cell cycle. From their studies, they determined that silver does not allow cancer cells to enter G1 phase. In the cell cycle, G1 phase is defined as the growth phase of the cycle where many biosynthetic pathways, including synthesis of the enzymes needed for DNA replication, resume [373]. It is an important phase in cell growth and may have a major link to why silver is active in cancer cells.

Silver has also recognized anti-bacterial activity. In the years, silver nitrate solution was known to have anti-bacterial properties. Au(II) compounds are nowadays attractive as bactericides. They have good solubility in water, contrasting Au(I) inorganic salts (excluding $AgNO_3$ and AgF), though silver nanoparticles have been currently developed as a prevalent way of monitoring bacteria and viruses. Moreover, one of the most proficient water decontamination systems includes also silver constituents.

11.6 Platinum

Platinum complexes, particularly cisplatin, are the best known of the metal-based compounds for cancer treatment. Cisplatin was firstly synthesized in 1844 [374]. Its completely unexpected anti-tumour activity was discovered in 1965 by Barnett Rosenberg at the study if the rate of bacteria growth at the electric field conditions [375, 376]. Agreement by the Food and Drug Administration for the [*cis*-diamminedichloridoplatinum(II)] as a drug for treatment of cancer has been given later in 1978 [377]. The complex is nowadays existing for treatment and is known as cisplatin. The reason for the compound's efficiency appears to be the capability of the *cis*-$(H_3N)_2Pt$ molecule to cross-link DNA parts, thus avoiding additional synthesis of DNA.

However, resistance to cisplatin has been occurred. Nevertheless, it becomes the essential basis which activated the search for alternate metal-based compounds with enhanced anti-neoplastic and pharmacokinetic activities. On this base, thousands of alternate Pt-based compounds have been developed, similar to the parent cisplatin. Carboplatin, oxaliplatin, ormaplatin, satraplatin, aroplatin, sebriplatin, enloplatin, miboplatin, zeniplatin, satraplatin, picoplatin and iproplatin – all of them are products of wide-ranging study of platinum coordination compounds. Till now, relatively few of them finalized the clinical trials [378] and six are presently approved, specifically – cisplatin [*cis*-diamminedichloridoplatinum(II), $PtCl_2(NH_3)_2$], oxaliplatin [(*R*,*R*)-diaminocyclohexane-1,2-ethanedicarboxylatoplatinum(II)], nedaplatin [*cis*-diammine-2-hydroxyacetatoplatinum(II); Japan], carboplatin [*cis*-diammine-1, 1-cyclobutanedicarboxylatoplatinum(II)], lobaplatin [*cis*-1,2-diamminocyclobutane-2-hydroxypropanoatoplatinum(II); China] and heptaplatin [*cis*-malonato-[(4*R*,5*R*)-4,5-*bis*(aminomethyl)-2-isopropyl-1,3-dioxolane]platinum(II); South Korea]. After preliminary achievement of the discovery of cisplatin, the second-generation Pt drug (carboplatin) shows lacking of nephrotoxicity accompanied by decreased gastrointestinal harmfulness. The third-generation Pt drugs involve oxaliplatin, which also disables the resistance and is active in the treatment of testicular and ovarian cancer. Heptaplatin, or otherwise known as SKI-2053 R, is used for treatment of gastric cancer [379–382]. It is permitted for therapy combined with 5-fluorouracil and the compound showed reduced nephrotoxicity in comparison to the parent cisplatin [383, 384].

11.6.1 Cisplatin

Presently, the anti-neoplastic cisplatin is endorsed and is known as *cisplatinol®* and *platinosin®*. Cisplatin has been permitted for the treatment of ovarian, lung testicular, head, neck tumours, as well as breast metastatic cancer [385]. This platinum drug has shown response rates between 70% and 90% in these types of tumours with survival rates ranging anywhere from 12 to 38 months depending on the cancer type and severity, as shown in Table 11.1

Table 11.1: Response rates and survival time for cancers treated by cisplatin.

Tumour type	Response rate (%)	Median survival rate (months)
Small cell lung cancer	80–95	12–16
Ovarian (stages II–IV)	70–80	26–38
Testicular (stage I)	80–99	Most patients cured
Head and neck	80–90	12–36
Non-small-cell lung cancer	20	12

Cisplatin displays enhanced therapeutic rate for the treatment of testicular cancer and for ovarian cancer [386]. It has to be mentioned that only the *cis* geometric isomer of [diamminedichloridoplatinum(II), $PtCl_2(NH_3)_2$] was found active. The *trans* isomer showed no pharmacological activity. Therefore, these coordination compounds illustrate the isomerism impact on the chemical performance. There are some key problems of the usage of cisplatin for tumour treatment. These limitations are its low saline solubility, resistance and serious adverse effects [387]. Additionally, cisplatin has to be applied intravenously, which is rather problematic. Colorectal tumours and non-small-cell lung cancers show essential resistance to the drug cisplatin [388]. This resistance is generally multi-factorial [389].

11.6.2 Second-generation platinum(II) drugs

A structure–activity relationship (SAR) has been developed in platinum anti-neoplastic research field [390, 391]. These basic guidelines help to project new drugs comparable to cisplatin activity. According to the SAR, it has been proven that Pt(+2) or Pt(+4) complexes should possess *cis*-geometrical conformation with the formula *cis*-$[PtX_2(Am)_2]$ or *cis*-$[PtX_2Y_2(Am)_2]$ (X = leaving functional group; Am = passive amine). The respective amine group should have a minimum one NH moiety. The requirement for the leaving functional groups X are to be anions with middle binding to Pt and a weak *trans*-effect to stop Am release in biological environment. It has to be taken into consideration that compounds with non-stable leaving groups, for example ClO_4^- and NO_3^-, are extremely poisonous, middle labile Cl^- or Br^- are anti-tumour potent, while complexes with comparatively passive leaving group, for instance I^-, N_3^- or SCN^-, are not active to tumours and are non-cytotoxic. With the purpose of overcoming the limitations exerted by cisplatin, for example toxicity, acquired and inherent resistance as well as low saline solubility, second-generation medications were established. These metal-based complexes are composed of less-labile ions like oxalate, carboxylate or glycolate. The most promising compound within the group is the drug carboplatin that holds the multidentate cyclobutanedicarboxylic acid. It displays better therapy index and overcomes some of the observed adverse effects. Carboplatin is used in the combination treatment for the management of ovarian cancer. This compound shows lower activity and lower toxicity than the parent cisplatin [392]. Lobaplatin is a platinum derivative, signified as 1,2-diammino-l-methyl-cyclobutane-platinum(II)-lactate. It displays equal or higher in vivo and in vitro potency related to the drugs carboplatin or cisplatin. Lobaplatin does not show cross-resistance to cisplatin [393, 394].

11.6.3 Third-generation platinum(II) drugs

The third-generation compounds contain various chiral amines [395] accompanied by carboxylate ions as ligands. Oxaliplatin is the most potent drug in the group. This is a Pt compound containing oxalate and diaminocyclohexane (DACH). DACH does not show cross-resistance to oxaliplatin and cisplatin and acts a chief role in cytotoxic activity. It is approved for the combination treatment with additional chemotherapeutics in the therapy of non-small-cell lung and colon cancers. The medication shows improved protection profile compared to cisplatin, and that is why it can be applied in patients not tolerating cisplatin. The second compound in this series is [*R*, *R*-1,2-diaminocyclohexane bis-neodecanoatoplatinum(II)], L-NDDP. This compound is still in active research [396–400]. L-NDDP is the first platinum compound with liposomal properties considered in clinical research [401]. The motivation for this alteration is to solubilize better the Pt complex and to decrease the nephrotoxicity, as well as its cross-resistance [402–404].

Regrettably, no one of the replacements of the drug cisplatin could be found better than the original compound in relation to toxic activity and medical effectiveness and, therefore, the search for innovative Pt-based complexes with optimized properties continues to be an unresolved problem. Many Pt-based compounds have been tested by clinical experiments in order to discover an alternate to *cis*-platinum, principally attributable to side effects or to resistance. The absence of greater benefit of the new preparations over cisplatin led to the postponement of the research. Pt drugs suspended from clinical tests involve sebriplatin, cycloplatam, spiroplatin, aroplatin, miboplatin, SPI-077, BBR3464, zeniplatin, iproplatin, ormaplatin JM 11, enloplatin and NSC 170898. TRK-710 is an alternative referent compound, which was considered in phase-I clinical tests in Japan. The non-appearance of cisplatin cross-resistance observed in both in vitro and in vivo predominantly in the L1210/CDDP model [CDDP = *cis*-diamminedichloridoplatinum(II)] decreased the toxicity, and the observed different mode of action stimulated the clinical progress. The other complexes of the DACH series are not established further due to the adverse effects, poisonousness and chemical lability [405, 406].

An entirely different line to dynamic drug design is to include ligands showing steric crowding. These compounds do not achieve the standards of third-generation medications, though its inclusion in the group of non-classical Pt drugs is not achievable. A well-known compound is picoplatin. The compound is a 2-methylpyridine analogue of cisplatin (previously identified as ZD0473) firstly established to afford steric protection round the Pt centre, thus giving a steric interference to the medication and avoiding the nucleophilic attack. Preclinical research revealed promising anti-neoplastic activities in the resistant to cisplatin cell lines. Such compounds violate the SAR rules of cisplatin-corresponding complexes but show substantial anti-tumour potency. Picoplatin was proven to be active in the administration of ovarian tumour cell lines resistant to carboplatin and cisplatin [407–409]. Picoplatin has confirmed its activity in various solid tumours, involving ovarian, lung, colorectal and prostate cancers [410].

11.6.4 Non-classical platinum compounds

Recently, a wide variety of non-classical Pt compounds have been obtained, and their anti-neoplastic activities against tumour cells have been estimated. This includes modifying the parent compounds by using different ligands to reach the anticipated results. Generally, the respective pharmacokinetic properties, range of action and cytotoxicity are enhanced to avoid those problems characteristic for the parent complexes. They include *trans*-Pt(II) compounds [411], platinum(IV) compounds, sterically hindered Pt compounds, Pt(II) and Pt(IV) complexes with pharmacologically appropriate carrier ligands, Pt complexes with intercalator organic ligands, as well as multifunctional polynuclear Pt complexes. The typical feature of these complexes is a definitely dissimilar mechanism of action to living targets such as DNA or cellular protein. Consequently, they were anticipated to overcome the resistance.

11.6.5 Trans-platinum(II) compounds

It is well known that the *trans* isomer of the drug cisplatin 'transplatin' demonstrates no anti-tumour potential. Consequently, consistent with SAR rules the *trans* complexes would not be supposed to show activity. Nevertheless, Farrell et al. [412] created a group of potent *trans* complexes with a key formula, *trans*-[$PtCl_2(L)(NH_3)$] (L = heterocyclic amine in planar conformation). The *trans-E,E*-iminoether platinum(II) compound [413, 414] exhibited a noticeable improvement in cytotoxic activity compared to inactive *trans*-platin and significant anti-tumour activity resistant to cisplatin cancer cells. Other *trans* compounds have also been studied [415–418].

11.6.6 Platinum(IV) compounds

Innovative methods have been implemented to explore other obtainable Pt-related compounds. These are centred around Pt(IV) compounds, whose anti-neoplastic activity has been documented during the last decades. Their stability and extended sphere assist as a benefit of overcoming the problems typical for Pt(II) complexes. Some platinum(IV) complexes have been examined as orally potent medications. They are suitable for intestinal absorption because they have lipophilic groups at axial positions. Platinum(IV) complexes can be reduced to active Pt(II) prior to reaction with DNA, by extra- and intracellular reducing agents, so they are typical 'prodrugs'. They have six saturated coordination spheres with octahedral geometric conformation. This feature provides the kinetic stability over the Pt(II) compounds. The most promising one is [bis-(acetato)-amminedichlorido(cyclohexylamine)platinum(IV)], the so-called satraplatin or JM216, which is the first orally administrated bioavailable Pt medication. This drug displays variable pharmacodynamic and kinetic properties comparative to other Pt complexes and henceforth may have a diverse range of anti-neoplastic activity. Two additional effective complexes are JM335 [*trans,cis, trans*-amminedichlorido(cyclohexylamine)dihydroxidoplatinum(IV)] and ormaplatin [*cis*-tetrachlorido(1,2-cyclohexyldiamine)platinum(IV)] [419].

11.6.7 Platinum compounds with intercalator ligands

These complexes have structures which are diverse from that of cisplatin. They have aromatic heterocyclic intercalators in planar geometry, for instance terpyridine, bipyridine, phenanthroline [420]. Pt(II) terpyridine complexes are the best considered as Pt intercalators. $[Pt(terpy)(SC_2H_4OH)]^+$ shows intercalating DNA binding [421]. The usage of a fourth ligand, like neutral 4-picoline, produces 2+ ions, with better water solubility. Some of the mentioned complexes were experienced against L1210 leukaemic cell line in vitro and in vivo. The complex ion $[Pt(terpy)Cl]^+$ has not displayed potency though the other terpy-Pt complexes have been observed to be active [422].

11.6.8 Polynuclear platinum compounds

Polynuclear Pt coordination complexes represent a distinct structural group of DNA-binding candidates with excellent anti-neoplastic properties. The use of as a minimum two Pt coordinating units directly means that multifunctional DNA binding approaches are possible. The strategy of novel anti-tumour compounds is focused towards the di-nuclear and tri-nuclear compounds [423–430]. They can interact with DNA by making long-variety of interstrand and intrastrand cross-links at several sites [431]. Wide choice is possible to vary the metal-coordination sphere: leaving groups, amine ligands and backbone linkers. The di-nuclear motif involving two *cis*-$PtCl(NH_3)_2$ parts connected through a flexible diamine-linker [432]. The complex $[Cl(NH_3)_2PtH_2N(CH_2)nNH_2PtCl(NH_3)_2]$ ($n = 4$) may occur in different forms of isomers: covering the coordination parts in *cis* conformation (2,2/*c*,*c*), both coordination units in *trans* (2,2/*t*,*t*) or in mixed *cis*,*trans* (2,2/*c*,*t*) types [433]. SAR concerning the chain-length and anti-tumour capacity is detected for the better potent *trans*-isomer [434]. It has been observed that tri-functional binuclear complexes interact with DNA and successfully produce inter-strand cross-link to DNA [435]. Cytotoxic profiles of these complexes have been investigated on ovarian cancer cells [436] and on the resistant cell line 41 M/*cis*R [437].

Triplatin complex BBR3464 has experienced phase II clinical test for administration on various malignancies. In the complex, three Pt coordination components are associated with polyamine linkers. The high oxidation state (4+) accompanied by H-bonding potential makes this complex highly vulnerable for electrostatic affinity to DNA [438–442].

The polyamine complexes of spermidine and spermine have polycationic profile and become protonated at biological values of pH. The linkers may afford non-covalent contacts with the negative DNA backbone [443–447]. The Pt complexes with spermidine and spermine demonstrated good cytotoxic properties in cells like L1210/0 and in the resistant cell line L1210/CDDP. Other similar substituted compounds with less toxic effects and better acceptance have been synthesized and examined for anti-neoplastic activity. These so-called blocked polyamines after coordination with platinum have less cytotoxicity than the separate ligand spermidine and consequently can be improved medicines [448, 449].

Many Pt drugs are still mostly effective against the same class of cancers that cisplatin is active against; however, they have shown efficacy against tumours that have acquired cisplatin resistance [450]. Still, the same toxicity factors usually arise with the use of these new derivatives as shown in Table 11.2. Though the use of cisplatin and its derivatives has led to a major improvement in the treatment of different tumours, the limitations due to the possible toxic effects and resistance have encouraged major research efforts into the development of numerous non-platinum -based chemotherapeutics for the treatment of cancer.

Table 11.2: Toxicity of cisplatin and its analogues.

Drug	Max. dose (mg/kg)	Major limiting toxicity
Cisplatin	3	Nephrotoxicity[a]
Carboplatin	30	Nephrotoxicity[a]
Nedaplatin	3	Myelosuppression[b]
Cycloplatam	3	Myelosuppression[b]
SKI 2053 R	10	Hepatotoxicity[c]

[a] Severe toxicity to the kidneys.
[b] A condition where bone marrow activity decreases, resulting in less red and white blood cells, as well as platelets.
[c] Severe toxicity to the liver.

References

[1] Roat-Malone, R. M., Bioinorganic Chemistry – A Short Course, Washington College, Chestertown, MD, John Wiley & Sons, Inc., Hoboken, New Jersey, 2003.

[2] Medvetz, D. A., The synthesis, characterization, and antitumor properties of Ag(I), Cu(II), and Rh(III) metal complexes. Dissertation, University of Akron, 2008.

[3] McGowan, P. C., Metal complexes as pharmaceuticals, Annu. Rep. Prog. Chem., Sect. A, 2005, 101, 631–648.

[4] Mudgal, V., N., Madan, A., Mudgal, R., Singh, S., Mishra, The Open Nutraceut. J., 3, 2010, 94.

[5] http://www.uv.es/EBRIT/macro/macro_5001_20_267.html

[6] Rehder, D., Introduction to Bioinorganic Chemistry, Lecture notes, University of Lund, 2008.

[7] Kienle, L., Basics of Bioinorganic Chemistry, Max-Planck-Institut für Festkörperforschung, Stuttgart, 2008.

[8] Blake, B. L., C. P., Lawler, R. B., Mailman, Introduction to Biochemical Toxicology, 3rd, Hodgson, E., Smart, R. C., Eds., Wiley Interscience, New York, 2001.

[9] Ermidou-Pollet, S., S., Pollet, International Symposium on Trace Element in the Food Chain, 2006, 357–361.

[10] Sharma, P., S., Johri, S., Shukla, J. Appl. Toxicol., 20, 2000, 313.

[11] Roveda, L., U., Prati, J., Bakeine, F., Trotta, P., Marotta, P., Valsecchi, A., Zonta, R., Nano, A., Facoetti, P., Chiari, S., Barni, T., Pinelli, S., Altieri, P., Pedroni, J. Chemother., 16, 2004, 15.

[12] Miura, M., G. M., Morris, P. L., Micca, M. M., Nawrocky, M. S., Makar, S. P., Cook, D. N., Slatkin, Br. J. Radiol., 77, 2004, 573.

[13] Nakamura, H., Y., Miyajima, S., Kasaoka, K., Maruyama, Chem. Commun., 2004, 1910.

[14] Lee, C.-H., I. D., Yang, J.-D., Lee, H., Nakamura, J., Ko, S. O., Kang, Synlett, 2004, 1799.

[15] Collery, P., B., Keppler, C., Madoulet, B., Desoize, Crit. Rev. Oncol. Hemat., 42, 2002, 283.

[16] Bernstein, L. R., Pharmacol. Rev., 50, 1998, 665.

[17] Bernstein, L. R., Therapeutic Gallium Compounds, Wiley, New York, 2005.

[18] Jakupec, M. A., B. K., Keppler, Curr. Top. Med. Chem., 4, 2004, 1575.

[19] Chitambar, C. R., P. A., Seligman, J. Clin. Invest., 78, 1986, 1538.

[20] Chitambar, C. R., Z., Zivkovic, Cancer Res., 47, 1987, 3929.

[21] Kinuya, S., X. F., Li, K., Yokoyama, H., Mori, K., Shiba, N., Watanabe, N., Shuke, H., Bunko, T., Michigishi, N., Tonami, Nucl. Med. Commun., 25, 2004, 49.

[22] Valiahdi, S. M., M. A., Jakupec, R., Marculescu, W., Berger, K., Rappersberger, B. K., Keppler, Mol. Cancer Ther, 6, 2007, 3426S.

[23] Perchellet, E. M., J. B., Ladesich, P., Collery, J. P., Perchellet, Anti-Cancer Drugs, 10, 1999, 477.

[24] Niesvizky, R., Semin. Oncol., 30, 2003, 20.

[25] Chitambar, C. R., Curr. Opin. Oncol, 16, 2004, 547.

[26] Jakupec, M. A., M., Galanski, V., Arion, C., Hartinger, B. K., Keppler, Dalton Trans., 2008, 183.

[27] Kowol, C. R., R., Berger, R., Eichinger, A., Roller, M. A., Jakupec, P. P., Schmidt, V. B., Arion, B. K., Keppler, J. Med. Chem, 50, 2007, 1254.

[28] Rudnev, A. V., L. S., Foteeva, C., Kowol, R., Berger, M. A., Jakupec, V. B., Arion, A. R., Timerbaev, B. K., Keppler, J. Inorg. Biochem., 100, 2006, 1819.

[29] Timerbaev, A. R., O. O., Vasylenko, L. S., Foteeva, A. V., Rudnev, O., Semenova, B. K., Keppler, J. Sep. Sci, 30, 2007, 399.

[30] Collery, P., F., Lechenault, A., Cazabat, E., Juvin, L., Khassanova, A., Evangelou, B., Keppler, Anticancer Res., 20, 2000, 955.

[31] Chua, M. S., L. R., Bernstein, R., Li, S. K., So, Anticancer Res., 26, 2006, 1739.

[32] Lum, B. L., A., Gottlieb, R., Altman, P., Sayre, F., Valone, J. Bone Miner. Res, 18, 2003, S391.

[33] Van Den Bossche, B. C., C., Van De Wiele, J. Clin. Oncol., 22, 2004, 3593.

https://doi.org/10.1515/9781501516115-016

[34] Onthank, D. C., S., Liu, P. J., Silva, J. A., Barrett, T. D., Harris, S. P., Robinson, D. S., Edwards, Bioconjugate Chem, 15, 2004, 235.
[35] Lei, V., W. K. A., Amoa-Awua, L., Brimer, Int. J. Food Microbiol, 53, 1999, 169.
[36] Gandini, C., J. Toxicol. Clin. Toxicol, 39, 2001, 35.
[37] Modifying factors of toxic effects, Basic Toxicology, Lu, F. C., Ed., Hemisphere Publishing Corp, New York, 1991, chap. 5, 61–73.
[38] Nakashima, H., J. Occup. Health, 40, 1998, 270.
[39] Jain, M., S., Maanju, R. V., Singh, Appl. Organomet. Chem, 18, 2004, 471.
[40] Jain, M., S., Gaur, S. C., Diwedi, S. C., Joshi, R. V., Singh, A., Bansal, Phosphorus, Sulfur Silicon Relat. Elem., 179, 2004, 1517.
[41] Jain, M., S., Gaur, V. P., Singh, R. V., Singh, Appl. Organomet. Chem., 18, 2004, 73.
[42] Imtiazud, D., M., Mazhar, K. M., Khan, K. C., Molloy, J. Organomet. Chem, 689, 2004, 899.
[43] Nath, M., R., Jairath, G., Eng, X., Song, A., Kumar, Inorg. Chem. Commun., 7, 2004, 1161.
[44] Sadiqur, R., S., Ali, A., Badshah, A., Malik, E., Ahmed, G.-X., Jin, E. R. T., Tiekink, Appl. Organomet. Chem, 18, 2004, 401.
[45] Sadiqur, R., S., Ali, A., Badshah, M., Mazhar, X., Song, G., Eng, K. M., Khan, Synth. React. Inorg. and Met.-Org. Chem., 34, 2004, 1379.
[46] Ashfaq, M., M. I., Khan, M. K., Baloch, A., Malik, J. Organomet. Chem, 689, 2004, 238.
[47] Chaudhary, A., A., Phor, S., Gaur, R. V., Singh, Heterocycl. Commun, 10, 2004, 181.
[48] Chaudhary, A., A., Phor, G. K., Agarwal, R. V., Singh, Heterocycl. Commun., 10, 2004, 393.
[49] Kaste, J. M., A. J., Friedland, S., Sturup, Environ. Sci. Technol, 37, 2003, 3560.
[50] Tandon, S. K., M., Chatterjee, A., Bhargava, V., Bihari, Sci. Total. Environ., 281, 2001, 177.
[51] Siddiqui, M. K., S., Srivastava, P. K., Mehrotra, Biomed. Environ. Sci, 15, 2002, 298.
[52] Lindbohm, M. L., M., Sallmen, A., Anttila, H., Taskinen, K., Hemminki, Scand. J. Work Environ. Health, 17, 1991, 95.
[53] Hernberg, S., Am. J. Industrial Med, 38, 2000, 244.
[54] Goyer, R. A., Toxic effects of metals, Casarett and Doull's Toxicology: The Basic Science of Poisons, 5th, Klaassen, C. D., Ed., McGraw-Hill, New York, 1996, chap, Vol. 23, 691.
[55] Quistad, G. B., Chem. Res. Toxicol, 13, 2000, 652.
[56] Tchounwou, P. B., J. A., Centeno, A. K., Patlolla, Mol. Cell. Biochem., 255, 2004, 47.
[57] Wanibuchi, H., E. I., Salim, A., Kinoshita, J., Shen, M., Wei, K., Morimura, K., Yoshida, K., Kuroda, G., Endo, S., Fukushima, Toxicol. Appl. Pharmacol., 198, 2004, 366.
[58] Liu, Z., E., Boles, B. P., Rosen, J. Biol. Chem, 279, 2004, 17312.
[59] Yamauchi, H., Y., Aminaka, K., Yoshida, G., Sun, J., Pi, M. P., Waalkes, Toxicol. Appl. Pharmacol, 198, 2004, 291.
[60] Apostoli, P., Fresenius J. Anal. Chem., 363, 1999, 499.
[61] Yamato, N., Bull. Environ. Contamin. Toxicol., 40, 1988, 633.
[62] Michel, P., Arsenic in marine medium; biochemistry and ecotoxicology, Pepères, Ocean n 4th ed. Ifremer, Paris, 1993.
[63] Ui, J., Industrial pollution in Japan, The United Nations University, Tokyo, 1992.
[64] Lee, M. Y., B. I., Jung, S. M., Chung, Environ. Health. Perspect, 111, 2003, 513.
[65] Yoshida, T., H., Yamauchi, G., Fan Sun, Toxicol. Appl. Pharmacol, 198, 2004, 243.
[66] Sarkar, M., G. R., Chaudhuri, A., Chattopadhyay, N. M., Biswas, Asian J. Androl., 5, 2003, 27.
[67] Walton, F. S., A. W., Harmon, D. S., Paul, Z., Drobná, Y. M., Patel, M., Styblo, Toxicol. Appl. Pharmacol, 198, 2004, 424.
[68] Feldmann, J., Spec. Publ. R. Soc. Chem., 267, 2001, 380.
[69] Yao, Z., M., Zhang, K., Garmestani, D. B., Axworthy, R. W., Mallett, A. R., Fritzberg, L. J., Theodore, P. S., Plascjak, W. C., Eckelman, J. A., Carrasquillo, Clin. Cancer Res., 10, 2004, 3137.

[70] Yu, L., Y.-Q., Ma, G.-C., Wang, H.-B., Song, H.-G., Wang, J.-S., Li, J.-R., Cui, R.-Q., Wang, Appl. Organomet. Chem., 18, 2004, 187.
[71] Nomiya, K., K., Sekino, M., Ishikawa, A., Honda, M., Yokoyama, K. N., Chikaraishi, H., Yokoyama, S., Nakano, K., Onodera, J. Inorg. Biochem, 98, 2004, 601.
[72] Bland, M., S., Ismail, J., Heinemann, J., Keenan, Antimicr. Agents Chemother., 48, 2004, 1983.
[73] Milenic, D. E., K., Garmestani, E. D., Brady, P. S., Albert, D., Ma, A., Abdulla, M. W., Brechbiel, Clin. Cancer Res, 10, 2004, 7834.
[74] Furne, J., Biochem. Pharmacol., 62, 2001, 255.
[75] Smith, R. P., Toxic responses of the blood, Casarett and Doull's Toxicology: The Basic Science of Poisons, 5th, Ed., McGraw-Hill, New York, 1996, chap. 11, 335–354.
[76] Costa, D. L., Air pollution, Casarett and Doull's Toxicology: The Basic Science of Poisons, 5th, Klaassen, C. D., Ed., McGraw-Hill, New York, 1996, chap. 28, 857–882.
[77] Trenga, C. A., J. Q., Koenig, P. V., Williams, Occup. Environ. Med, 56, 1999, 544.
[78] Newhook, R., M. E., Meek, M., Walker, Environ. Carcinog. Ecotoxicol. Rev., C19, 2001, 125.
[79] Mézes, M., K., Balogh, Selenium Supplimentation in animal and man – positive effects and negative consequences. Intern. Symposium on Trace Element in the Food Chain, 2006; 9–15.
[80] Patrick, L., Altern. Med. Rev., 9, 2004, 239.
[81] Felix, K., S., Gerstmeier, A., Kyriakopoulos, H.-F., Dong, M., Eckhaus, D., Behne, G. W., Bornkamm, S., Janz, Cancer Res, 64, 2004, 2910.
[82] Jonsson-Videsater, K., L., Bjorkhem-Bergman, A., Hossain, A., Soderberg, L. C., Eriksson, C., Paul, A., Rosen, M., Bjornstedt, Biochem. Pharmacol., 67, 2004, 513.
[83] Zhang, J., H., Wang, Y., Bao, L., Zhang, Life Sci, 75, 2004, 237.
[84] Lund, K., Occup. Environ. Med, 54, 1997, 32.
[85] Ueno, H., Y., Sayato, K., Nakamuro, J. Health Sci., 46, 2000, 110.
[86] Orlova, A., A., Sjoestroem, O., Lebeda, H., Lundqvist, J., Carlsson, V., Tolmachev, Anticancer. Res, 24, 2004, 4035.
[87] Wesley, J. N., E. C., McGee, K., Garmestani, M. W., Brechbiel, A. T., Yordanov, C., Wu, O. A., Gansow, W. C., Eckelman, J. D., Bacher, M., Flynn, C. K., Goldman, M., MacLin, U. P., Schwartz, T., Jackson-White, C. M., Phillip, J., Decker, T. A., Waldmann, Nucl. Med. Biol, 31, 2004, 357.
[88] Yordanov, A. T., O., Pozzi, S., Carlin, G., Akabani, B., Wieland, M. R., Zalutsky, J. Radioanal Nucl. Chem., 262, 2004, 593.
[89] Wilbur, D. S., M. K., Chyan, D. K., Hamlin, B. B., Kegley, R., Risler, P. M., Pathare, J., Quinn, C., Foulon, M., Zalutsky, T. J., Wedge, M. F., Hawthorne, Bioconjugate Chem, 15, 2004, 203.
[90] Frausto Da Silva, J. R., J. P., Williams, The Biological Chemistry of the Elements: The Inorganic Chemistry of Life, Clarendon Press, New York, 1991.
[91] Lippard, S. J., J. M., Berg, Principles of Bioinorganic Chemistry, University Science Books, Mill Valley, CA, 1994.
[92] Hay, R. W., Bio-Inorganic Chemistry, Ellis Horwood, Halsted Press, New York, 1984.
[93] Petering, D., Metal Ions Biol. Med. 11, 1980, 197.
[94] Gokhale, N., S., Padhye, S., Padhye, C., Anson, A., Powell, Inorg. Chim. Acta, 319, 2001, 90.
[95] Collins, M., D., Ewing, G., Mackenzie, E., Sinn, U., Sandbhor, S., Padhye, S., Padhye, Inorg. Chem. Comm, 3, 2000, 453.
[96] Zou, G., X., Cai, N., Pan, Biometals, 20, 2007, 1.
[97] Somasundaram, K., J. Med. Chem, 48, 2005, 977.
[98] Butte, T., Immun. Today., 15, 1994, 7.
[99] Levine, A., Cell., 88, 1997, 323.
[100] Safaei, R., A. K., Holzer, K., Katano, S. B., Howell, J. Inorg. Biochem., 98, 2004, 1607.

[101] Guo, Y., K., Smith, M. J., Petris, J. Biol. Chem, 279, 2004, 46393.
[102] Szczepanik, W., P., Kaczmarek, M., Jezowska-Bojczuk, J. Inorg. Biochem, 98, 2004, 2141.
[103] Szczepanik, W., A., Czarny, E., Zaczynska, M., Jezowska-Bojczuk, J. Inorg. Biochem, 98, 2004, 245.
[104] Zhang, S., Y., Zhu, C., Tu, H., Wei, Z., Yang, L., Lin, J., Ding, J., Zhang, Z., Guo, J. Inorg. Biochem, 98, 2004, 2099.
[105] Chaviara, A. T., P. J., Cox, K. H., Repana, R. M., Papi, K. T., Papazisis, D., Zambouli, A. H., Kortsaris, D. A., Kyriakidis, C. A., Bolos, J. Inorg. Biochem, 98, 2004, 1271.
[106] Ye, N., G., Park, A. M., Przyborowska, P. E., Sloan, T., Clifford, C. B., Bauer, G. A., Broker, R. D., Rogers, R., Ma, S. V., Torti, M. W., Brechbiel, R. P., Planalp, Dalton Trans., 2004, 1304.
[107] Barragan, E., B., Gordillo, G., Vargas, L., Velazco, Appl. Organomet. Chem., 18, 2004, 311.
[108] Zhang, S., C., Tu, X., Wang, Z., Yang, J., Zhang, L., Lin, J., Ding, Z., Guo, Eur. J. Inorg. Chem., 2004, 4028.
[109] Devereux, M., M., McCann, D. O., Shea, R., Kelly, D., Egan, C., Deegan, K., Kavanagh, V., McKee, G., Finn, J. Inorg. Biochem, 98, 2004, 1023.
[110] Ferrari, M. B., F., Bisceglie, G., Pelosi, P., Tarasconi, R., Albertini, P. P., Dall'Aglio, S., Pinelli, A., Bergamo, G., Sava, J. Inorg. Biochem, 98, 2004, 301.
[111] John, R. P., A., Sreekanth, V., Rajakannan, M. P., Kurup, Polyhedron, 23, 2004, 2549.
[112] Cowley, A. R., J. R., Dilworth, P. S., Donnelly, J. M., Heslop, Dalton Trans., 2004, 2404.
[113] Baldini, M., M., Belicchi-Ferrari, F., Bisceglie, P. P., Dall'Aglio, G., Pelosi, S., Pinelli, P., Tarasconi, Inorg. Chem., 43, 2004, 7170.
[114] Ramadan, S., T. W., Hambley, B. J., Kennedy, P. A., Lay, Inorg. Chem, 43, 2004, 2943.
[115] Di Valra, M., C., Bazzicalupi, P., Orioli, L., Messori, B., Bruni, P., Zatta, Inorg. Chem., 43, 2004, 3795.
[116] Treiber, C., A., Simons, M., Strauss, M., Hafner, R., Cappaim, T. A., Bayer, G., Multhaup, J. Biol. Chem., 279, 2004, 51958.
[117] Segla, P., D., Miklos, P., Olejnikova, B., Kalinakova, D., Hudecova, M., Palicova, J., Svorec, M., Valko, M., Melnik, T., Glowiak, Inorg. Chim. Acta, 357, 2004, 4172.
[118] Roy, T. G., S., Hazari, B. K., Dey, C., Bader, D., Rehder, Eur. J. Inorg. Chem, 2004, 4115.
[119] Chohan, Z. H., K. M., Khan, C. T., Supuran, Appl. Organomet. Chem, 18, 2004, 305.
[120] Patel, R. N., N., Singh, K. K., Shukla, U. K., Chauhan, J., Niclos-Gutierrez, A., Castineiras, Inorg. Chim. Acta, 357, 2004, 2469.
[121] Boswell, C. A., X., Sun, W., Niu, G. R., Weisman, E. H., Wong, A. L., Rheingold, C. J., Anderson, J. Med. Chem, 47, 2004, 1465.
[122] Sun, X., J., Kim, A. E., Martell, M. J., Welch, C. J., Anderson, Nucl. Med. Biol, 31, 2004, 1051.
[123] Kalinowski, D., D., Richardson, Pharmacol. Rev., 57, 2005, 547.
[124] Aouad, F., A., Florence, Y., Zhang, F., Collins, C., Henry, Inorg. Chim. Acta, 339, 2002, 470.
[125] Richardson, D., Curr. Med. Chem., 12, 2005, 2711.
[126] Blatt, J., S., Stitely, Cancer Res., 47, 1987, 1749.
[127] Kalinowski, D., Y., Yu, P., Sharpe, M., Islam, Y., Liao, D., Lovejoy, N., Kumar, P., Bernhardt, D., Richardson, J. Med. Chem., 50, 2007, 3716.
[128] Asayama, S., N., Kasugai, S., Kubota, S., Nagaoka, H., Kawakami, J. Inorg. Biochem., 101, 2007, 261.
[129] Buss, J., B., Greene, J., Turner, F., Torti, S., Torti, Curr. Top. Med. Chem., 4, 2004, 1623.
[130] Stevens, R., D., Jones, M., Micozzi, P., Taylor, New Eng. J. Med, 319, 1988, 1047.
[131] Bulte, J. M., D. L., Kraitchman, Curr. Pharm. Biotechn, 5, 2004, 567.
[132] Nitin, N., L. E., LaConte, O., Zurkiya, G., Bao, JBIC, J. Biol. Inorg. Chem., 9, 2004, 706.
[133] Pierre, V. C., M., Melchior, D. M. J., Doble, K. N., Raymond, Inorg. Chem., 43, 2004, 8520.

[134] Bukowski, M. R., S., Zhu, K. D., Koehntop, W. W., Brennessel, L., Que Jr., JBIC, J. Biol. Inorg. Chem., 9, 2004, 39.
[135] Smolentsev, G., A. V., Soldatov, E. C., Wasinger, E. I., Solomon, Inorg. Chem., 43, 2004, 1825.
[136] Rajani, C., J. R., Kincaid, D. H., Petering, J. Am. Chem. Soc., 126, 2004, 3829.
[137] Domingo, J. L., Reprod. Toxicol., 10, 1996, 175.
[138] Goldwaser, I., J. Inorg. Biochem, 80, 2000, 21.
[139] Noblia, P., E. J., Baran, L., Otero, P., Draper, H., Cerecetto, M., Gonzalez, O. E., Piro, E. E., Castellano, T., Inohara, Y., Adachi, H., Sakurai, D., Gambino, Eur. J. Inorg. Chem., 2004, 322.
[140] Thompson, K. H., J. H., McNeill, C., Orvig, Chem. Rev, 99, 1999, 2561.
[141] Tracy, A. S., D. C., Crans, Vanadium Compounds: Chemistry, Biochemistry, and Therapeutic Applications., ACS Symposium Series 711, Oxford University Press, New York, 1998.
[142] Wilsky, G. R., A. B., Goldfine, P. J., Kostyniak, J. H., McNeill, L. Q., Yang, H. R., Khan, D. C., Crans, J. Inorg. Biochem, 85, 2001, 33.
[143] Reul, B. A., S. S., Amin, J. P., Buchet, L. N., Ongemba, D. C., Crans, S. M., Brichard, Br. J. Pharmacol., 126, 1999, 467.
[144] Amin, S. S., K., Cryer, B., Zhang, S. K., Dutta, S. S., Eaton, O. P., Anderson, S. M., Miller, B. A., Reul, S. M., Brichard, D. C., Crans, Inorg. Chem., 39, 2000, 406.
[145] Posner, B. I., R., Faure, J. W., Burgess, A. P., Bevan, D., Lachance, G., Zhang-Sun, I. G., Fantus, J. B., Ng, D. A., Hall, B., Soo Lum, A., Shaver, J. Biol. Chem, 269, 1994, 4596.
[146] Bevan, A. P., P. G., Drake, J., Yale, A., Shaver, B., Posner, Mol. Cell Biochem, 153, 1995, 49.
[147] Yale, J. F., C., Vigeant, C., Nardolillo, Q., Chu, J. Z., Yu, A., Shaver, B. I., Posner, Mol. Cell Biochem, 153, 1995, 181.
[148] Shaver, A., J. B., Ng, R. C., Hynes, B. I., Posner, Acta Crystallogr, C50, 1994, 1044.
[149] Yale, J. F., D., Lachance, A. P., Bevan, C., Vigeant, A., Shaver, B. I., Posner, Diabetes, 44, 1995, 1274.
[150] Melchior, M., S. J., Rettig, B. D., Liboiron, K. H., Thompson, V. G., Yuen, J. H., McNeill, C., Orvig, Inorg. Chem., 40, 2001, 4686.
[151] Mukherjee, B., B., Patra, S., Mahapatra, P., Banerjee, A., Tiwari, M., Chatterjee, Toxicol. Lett., 150, 2004, 135.
[152] Molinuevo, M. S., A. M., Cortizo, S. B., Etcheverry, Cancer Chemoth. Pharm., 61, 2008, 767.
[153] Upreti, R. K., Mol. Cell. Biochem, 153, 1995, 167.
[154] Thompson, H. J., N. D., Chasteen, L. D., Meeker, Carcinogenesis, 5, 1984, 849.
[155] Yang, X.-G., X.-D., Yang, L., Yuan, K., Wang, D. C., Crans, Pharm. Res., 21, 2004, 1026.
[156] Mukherjee, B., B., Patra, S., Mahapatra, M., Chatterjee, Toxicol. Lett., 150, 2004, 135.
[157] Osinska-Krolicka, I., H., Podsiadly, K., Bukietynska, M., Zemanek-Zboch, D., Nowak, K., Suchoszek-Lukaniuk, M., Malicka-Blaszkiewicz, J. Inorg. Biochem, 98, 2004, 2087.
[158] D'cruz, O. J., F. M., Uckun, J. Appl. Toxicol, 21, 2001, 331.
[159] Narla, R. K., C. L., Chen, Y. H., Dong, F. M., Uckun, Clin. Cancer Res., 7, 2001, 2124.
[160] Narla, R. K., Y. H., Dong, O. J., D'cruz, F. M., Uckun, Clin. Cancer Res., 6, 2000, 1546.
[161] Narla, R. K., Y. H., Dong, D., Klis, F. M., Uckun, Clin. Cancer Res., 7, 2001, 1094.
[162] D'cruz, O. J., F. M., Uckun, Expert Opin. Inv. Drug, 11, 2002, 1829.
[163] Dong, Y. H., R. K., Narla, E., Sudbeck, F. M., Uckun, J. Inorg. Biochem, 78, 2000, 321.
[164] Sun, W. Q., C. J., Weydert, L. W., Oberley, Free Radical Biol. Med., 35, 2003, S171.
[165] Lai, C. C., W. H., Huang, A., Askari, Y. H., Wang, N., Sarvazyan, L. M., Klevay, T. H., Chiu, Free Radical Biol. Med., 16, 1994, 613.
[166] Church, S. L., J. W., Grant, L. A., Ridnour, L. W., Oberley, P. E., Swanson, P. S., Meltzer, J. M., Trent, Proc. Nat. Acad. Sci. U.S.A., 90, 1993, 3113.

[167] Plymate, S. R., K. H., Haugk, C. C., Sprenger, P. S., Nelson, M. K., Tennant, Y. P., Zhang, L. W., Oberley, W. X., Zhong, R., Drivdahl, T. D., Oberley, Oncogene, 22, 2003, 1024.
[168] Decraene, D., L., Declercq, D., Gan, T., Mammone, D., Maes, M., Garmyn, J. Invest. Dermatol., 123, 2004, 207.
[169] Decraene, D., K., Smaers, D., Gan, T., Mammone, M., Matsui, D., Maes, L., Declercq, M., Garmyn, J. Invest. Dermatol., 122, 2004, 484.
[170] Niesel, J., A., Pinto, H. W. P., N'dongo, K., Merz, I., Ott, R., Gust, U., Schatzschneider, Chem. Commun., 2008, 1798.
[171] Wendland, M. F., NMR Biomed., 17, 2004, 581.
[172] Troughton, J. S., M. T., Greenfield, J. M., Greenwood, S., Dumas, A. J., Wiethoff, J., Wang, M., Spiller, T. J., McMurry, P., Caravan, Inorg. Chem., 43, 2004, 6313.
[173] Singh, R. V., A., Chaudhary, J. Inorg. Biochem., 98, 2004, 1712.
[174] Schaffer, A. W., J. Toxicol. Clin. Toxicol., 37, 1999, 839.
[175] Vincent, J. B., J. Nutr., 130, 2000, 715.
[176] Yamamoto, A., O., Wada, T., Ono, Eur. J. Biochem., 165, 1987, 627.
[177] Davis, C. M., J. B., Vincent, Biochemistry, 36, 1997, 4382.
[178] Davis, C. M., K. H., Sumrall, J. B., Vincent, Biochemistry, 35, 1996, 12963.
[179] Davis, C. M., A. C., Royer, J. B., Vincent, Inorg. Chem., 36, 1997, 5316.
[180] Tuzcu, A., M., Bahceci, M., Dursun, Y., Parmaksiz, M., Ertem, A., Dalgic, C., Turgut, E., Kale, J. Trace Elem. Exp. Med., 17, 2004, 55.
[181] Urmila, A. S., G., Sharma, J. X., Yan, S. D., Naranjan, K. G., Ramesh, Exp. Clin. Endocrinol. Diabetes, 112, 2004, 248.
[182] Pascal, L. E., D. M., Tessier, Toxicol. Lett., 147, 2004, 143.
[183] Pickering, A., C., Chang, J. B., Vincent, J. Inorg. Biochem., 98, 2004, 1303.
[184] Burger, R. M., Chem. Rev., 98, 1998, 1153.
[185] Böttcher, A., T., Takeuchi, K. I., Hardcastle, T. J., Meade, H. B., Gray, D., Cwikel, M., Kapon, Z., Dori, Inorg. Chem., 36, 1997, 2498.
[186] Takeuchi, T., A., Böttcher, C. M., Quezada, T. J., Meade, H. B., Gray, Bioorg. Med. Chem., 7, 1999, 815.
[187] Takeuchi, T., A., Böttcher, C. M., Quezada, M. I., Simon, T. J., Meade, H. B., Gray, J. Am. Chem. Soc., 120, 1998, 8555.
[188] Ji, L. N., X. H., Zou, J. G., Liu, Coord. Chem. Rev., 216, 2001, 513.
[189] Anderson, R. F., W. A., Denny, D. C., Ware, W. R., Wilson, Brit. J. Cancer, 74, 1996, S48.
[190] Ware, D. C., B. D., Palmer, W. R., Wilson, W. A., Denny, J. Med. Chem., 36, 1993, 1839.
[191] Ware, D. C., W., Wilson, W., Denny, C., Rickard, J. Chem. Soc., Chem. Commun., 1991, 1171.
[192] Gust, R., I., Ott, D., Posselt, K., Sommer, J. Med. Chem., 47, 2004, 5837.
[193] Chew, K.-B., M., Tarafder, K. A., Crouse, A. M., Ali, B. M., Yamin, H. K., Fun, Polyhedron, 23, 2004, 1385.
[194] Failes, T. W., C., Cullinane, C. I., Diakos, N., Yamamoto, J. G., Lyons, T. W., Hambley, Chem. Eur. J., 13, 2007, 2974.
[195] Ott, I., R., Gust, Arch. Pharm., 340, 2007, 117.
[196] Jung, M., D. E., Kerr, P. D., Senter, Arch. Pharm., 330, 1997, 173.
[197] Wiglenda, T., I., Ott, B., Kircher, P., Schumacher, D., Schuster, T., Langer, R., Gust, J. Med. Chem., 48, 2005, 6516.
[198] Ott, I., T., Koch, H., Shorafa, Z. L., Bai, D., Poeckel, D., Steinhilber, R., Gust, Org. Biomol. Chem., 3, 2005, 2282.
[199] Afrasiabi, Z., E., Sinn, J., Chen, Y., Ma, A. L., Rheingold, L. N., Zakharov, N., Rath, S., Padhye, Inorg. Chim. Acta, 357, 2004, 271.

[200] Rodriguez-Arguelles, M. C., M. B., Ferrari, F., Bisceglie, C., Pelizzi, G., Pelosi, S., Pinelli, M., Sassi, J. Inorg. Biochem., 98, 2004, 313.
[201] Leigh, G. J., Science, 268, 1995, 827.
[202] Dance, I., J. Biol. Inorg. Chem., 1, 1996, 581.
[203] Lehnert, N., B. E., Wiesler, F., Tuczek, D., Sellmann, J. Am. Chem. Soc., 119, 1997, 8869.
[204] Lehnert, N., B. E., Wiesler, F., Tuczek, D., Sellmann, J. Am. Chem. Soc., 119, 1997, 8879.
[205] Lehnert, N., F., Tuczek, Inorg. Chem., 38, 1999, 1671.
[206] Keppler, B. K., C., Friesen, H., Moritz, H., Vongerichten, E., Vogel, Struct. Bond., 78, 1991, 97.
[207] Harding, M. M., G., Mokdsi, Curr. Med. Chem., 7, 2000, 1289.
[208] Meléndez, E., Crit. Rev. Oncol. Hemat., 42, 2002, 309.
[209] Köpf, H., P., Köpfmaier, Angew. Chem. Int. Ed., 18, 1979, 477.
[210] Caruso, F., M., Rossi, Met. Ions Biol. Syst., 42, 2004, 353.
[211] Lummen, G., H., Sperling, H., Luboldt, T., Otto, H., Rubben, Cancer Chemoth. Pharm., 42, 1998, 415.
[212] Kroger, N., U. R., Kleeberg, K., Mross, L., Edler, D. K., Hossfeld, Onkologie, 23, 2000, 60.
[213] Guo, M. L., P. J., Sadler, J. Chem. Soc., Dalton Trans., 2000, 7.
[214] Guo, M. L., H. Z., Sun, H. J., McArdle, P. J., Sadler, Biochemistry, 39, 2000, 10023.
[215] Messori, L., P., Orioli, V., Banholzer, I., Pais, P., Zatta, FEBS Lett., 442, 1999, 157.
[216] Sun, H. Z., H. Y., Li, P. J., Sadler, Chem. Rev, 99, 1999, 2817.
[217] Yang, P., M. L., Guo, Coord. Chem. Rev., 185–6, 1999, 189.
[218] Fujishima, A., R. X., Cai, J., Otsuki, K., Hashimoto, K., Itoh, T., Yamashita, Y., Kubota, Electrochim. Acta, 38, 1993, 153.
[219] Huang, N. P., D., Huang, M. H., Xu, C. W., Yuan, Prog. Biochem. Biophys, 24, 1997, 470.
[220] Huang, N. P., M. H., Xu, C. W., Yuan, R. R., Yu, J. Photochem. Photobiol. A, 108, 1997, 229.
[221] Shavit, M., D., Peri, C. M., Manna, E. Y., Tshuva, J. Am. Chem. Soc., 129, 2007, 12098.
[222] Gansauer, A., I., Winkler, D., Worgull, T., Lauterbach, D., Franke, A., Selig, L., Wagner, A., Prokop, Chem. Eur. J., 14, 2008, 4160.
[223] Strohfeldt, K., M., Tacke, Chem. Soc. Rev., 37, 2008, 1174.
[224] Beckhove, P., O., Oberschmidt, A. R., Hanauske, C., Pampillon, V., Schirrmacher, N. J., Sweeney, K., Strohfeldt, M., Tacke, Anti-Cancer Drugs, 18, 2007, 311.
[225] Allen, O. R., L., Croll, A. L., Gott, R. J., Knox, P. C., McGowan, Organometallics, 23, 2004, 288.
[226] Potter, G. D., M. C., Baird, M., Chan, S. P. C., Cole, Inorg. Chem. Commun., 9, 2006, 1114.
[227] Maier, P., Cancer Chemother. Pharm., 23, 1989, 225.
[228] Valadares, M., A., Ramos, F., Rehmann, N., Sweeney, M., Strohfeldt, M., Quieroz, Euro. J. Pharm., 534, 2006, 264.
[229] Pampillon, C., N., Sweeney, K., Strohfeldt, M., Tacke, J. Organomet. Chem., 692, 2007, 2153.
[230] Melendez, E., Crit. Rev. Onc., 42, 2002, 309.
[231] Kopf-Maier, P., Naturwissenschaften., 68, 1981, 272.
[232] Rivera, C., Met. Ions Bio. Med., 6, 2000, 580.
[233] Causey, P. W., M. C., Baird, S. P. C., Cole, Organometallics, 23, 2004, 4486.
[234] Tacke, M., L. P., Cuffe, W. M., Gallagher, Y., Lou, O., Mendoza, H., Mueller-Bunz, F.-J. K., Rehmann, N., Sweeney, J. Inorg. Biochem., 98, 2004, 1987.
[235] Tacke, M., L. T., Allen, L., Cuffe, W. M., Gallagher, Y., Lou, O., Mendoza, H., Mueller-Bunz, F.-J. K., Rehmann, N., Sweeney, J. Organomet. Chem., 689, 2004, 2242.
[236] Sheikh, C., M. S., Hossain, M. S., Easmin, M. S., Islam, M., Rashid, Biol. Pharm. Bull., 27, 2004, 710.
[237] Kostova, I., Anti-Cancer Agents Med. Chem., 9, 2009, 827.
[238] Knochel, P., P., Jones, Eds., Organozinc Reagents: A Practical Approach, Oxford University Press, Oxford, 1999.

[239] Choi, C.-F., P.-T., Tsang, J.-D., Huang, E. Y. M., Chan, W.-H., Ko, W.-P., Fong, D. K. P., Ng, Chem. Commun., 2004, 2236.
[240] Tallkvist, J., C. L., Bowlus, B., Lonnerdal, Toxicol. Lett., 122, 2001, 171.
[241] Oudeh, M., M., Khan, J., Scullion, Environ. Pollut., 116, 2002, 293.
[242] Wu, X., T., Jin, Z., Wang, Q., Kong, G., Nordberg, J. Occup. Environ. Med., 43, 2001, 898.
[243] Henson, M. C., P. J., Chedrese, Exp. Biol. Med., 229, 2004, 383.
[244] Pongratz, R., K., Heumann, Anal. Chem., 68, 1996, 1262.
[245] Tokuomi, H., Y., Kinoshita, J., Teramoto, K., Imanishi, Hunter-Russel Syndrome. Nippon Rinsho, 35, 1977, 518.
[246] Kenntner, N., F., Tataruch, O., Krone, Environ. Toxicol. Chem., 20, 2001, 1831.
[247] Köpfmaier, P., Eur. J. Clin. Pharmacol, 47, 1994, 1.
[248] Metallotherapeutic drugs and metal-based diagnostic agents: the use of metals in medicine, John Wiley & Sons Ltd., Chichester, 2005.
[249] Messori, L., P., Orioli, G., Marcon, Biochem. Biophys. Res. Commun., 281, 2001, 352.
[250] Coronnello, M., G., Marcon, S., Carotti, B., Caciagli, E., Mini, T., Mazzei, P., Orioli, L., Messori, Oncol. Res., 12, 2001, 361.
[251] Suwalsky, M., P., Zambenedetti, E., Carpene, M., IbnLkayat, W., Wittkowski, L., Messori, P., Zatta, J. Inorg. Biochem., 98, 2004, 2080.
[252] Casini, A., C., Hartinger, C., Gabbiani, E., Mini, P. J., Dyson, B. K., Keppler, L., Messori, J. Inorg. Biochem., 102, 2008, 564.
[253] Messori, L., F., Abbate, G., Marcon, P., Orioli, M., Fontani, E., Mini, T., Mazzei, S., Carotti, T., O'connell, P., Zanello, J. Med. Chem., 43, 2000, 3541.
[254] Coronnello, M., E., Mini, B., Caciagli, M. A., Cinellu, A., Bindoli, C., Gabbiani, L., Messori, J. Med. Chem., 48, 2005, 6761.
[255] Messori, L., G., Marcon, M. A., Cinellu, M., Coronnello, E., Mini, C., Gabbiani, P., Orioli, Bioorg. Med. Chem., 12, 2004, 6039.
[256] Che, C. M., R. W. Y., Sun, W. Y., Yu, N. Y., Zhu, H. Z., Sun, Chem. Commun., 2003, 1718.
[257] Ronconi, L., L., Giovagnini, C., Marzano, F., Bettio, R., Graziani, G., Pilloni, D., Fregona, Inorg. Chem., 44, 2005, 1867.
[258] Ronconi, L., C., Marzano, P., Zanello, M., Corsini, G., Miolo, C., Macca, A., Trevisan, D., Fregona, J. Med. Chem., 49, 2006, 1648.
[259] Giovagnini, L., L., Ronconi, D., Aldinucci, D., Lorenzon, S., Sitran, D., Fregona, J. Med. Chem., 48, 2005, 1588.
[260] Buckley, R. G., A. M., Elsome, S. P., Fricker, G. R., Henderson, B. C., Theobald, R. V., Parish, B. P., Howe, L. R., Kelland, J. Med. Chem., 39, 1996, 5208.
[261] Tiekink, E. R. T., Crit. Rev. Oncol. Hemat., 42, 2002, 225.
[262] Mirabelli, C., R., Johnson, D., Hill, L., Faucette, G., Girard, J. Med. Chem., 29, 1986, 218.
[263] McKeage, M., L., Maharaj, S., Berners-Price, J. Coord. Chem., 232, 2002, 127.
[264] McKeage, M., S., Berners-Price, P., Galettis, W., Brouwer, L., Ding, L., Zhaung, B., Baguley, Cancer Chemother. Pharm., 46, 2000, 343.
[265] Racham, O., S., Nichols, S., Leedman, S., Berners-Price, A., Filipovska, Biochem. Pharm., 74, 2007, 992.
[266] Ortiz, A., H., Dulk, J., Brouwer, H., Kooijman, A., Spek, J., Reedijk, J. Inorg. Biochem., 101, 2007, 1922.
[267] Berners-Price, S., C., Mirabelli, R., Johnson, Cancer Res., 46, 1986, 486.
[268] McKeage, M., S., Berners-Price, L., Maharaj, Coord. Chem. Rev., 232, 2002, 127.
[269] Walz, D. T., M. J., Dimartino, D. E., Griswold, Scand. J. Rheumatol., 12, 1983, 16.
[270] Fries, J. F., Am. J. Med., 78, 1985, 12.
[271] Balkwill, F., A., Mantovani, Lancet, 357, 2001, 539.

[272] Lorenzen, I., Ann. Clin. Res., 7, 1975, 195.
[273] Ward, J. R., Am. J. Med., 85, 1988, 39.
[274] Simon, T. M., D. H., Kunishima, G. J., Vibert, A., Lorber, Cancer, 44, 1979, 44, 1965–1975.
[275] Simon, T. M., D. H., Kunishima, G. J., Vibert, A., Lorber, Cancer Res., 1981, 94.
[276] Mirabelli, C. K., R. K., Johnson, D. T., Hill, L. F., Faucette, G. R., Girard, G. Y., Kuo, C. M., Sung, S. T., Crooke, J. Med. Chem., 29, 1986, 218.
[277] Tiekink, E. R., P. D., Cookson, B. M., Linahan, L. K., Webster, Met.-Based Drugs, 1, 1994, 299.
[278] Webster, L. K., S., Rainone, E., Horn, E. R. T., Tiekink, Met.-Based Drugs, 3, 1996, 63.
[279] Berners-Price, S. J., C. K., Mirabelli, R. K., Johnson, M. R., Mattern, F. L., McCabe, L. F., Faucette, C. M., Sung, S. M., Mong, P. J., Sadler, S. T., Crooke, Cancer Res., 46, 1986, 5486.
[280] Hoke, G. D., G. F., Rush, C. K., Mirabelli, Toxicol. Appl. Pharmacol., 99, 1989, 50.
[281] Hoke, G. D., R. A., Macia, P. C., Meunier, P. J., Bugelski, C. K., Mirabelli, G. F., Rush, W. D., Matthews, Toxicol. Appl. Pharmacol., 100, 1989, 293.
[282] Barnard, P. J., M. V., Baker, S. J., Berners-Price, D. A., Day, J. Inorg. Biochem., 98, 2004, 1642.
[283] Rigobello, M. P., L., Messori, G., Marcon, M. A., Cinellu, M., Bragadin, A., Folda, G., Scutari, A., Bindoli, J. Inorg. Biochem., 98, 2004, 1634.
[284] Navarro, M., F., Vasquez, R. A., Sanchez-Delgado, H., Perez, V., Sinou, J., Schrevel, J. Med. Chem., 47, 2004, 5204.
[285] Kostova, I., Curr. Med. Chem. – Anti-Cancer Agents, 6, 2006, 19.
[286] Hughes, R., L., Bear, A., Kimball, Proc. Amer. Assoc. Cancer Res., 13, 1972, 120.
[287] Pruchnik, F., D., Dus, J. Inorg. Biochem., 61, 1996, 55.
[288] De Souza, A., R., Najjar, S., Glikmanas, S., Zyngier, J. Inorg. Biochem., 64, 1996, 1.
[289] Pruchnik, F., B., James, P., Kvintovics, Can. J. Chem., 64, 1986, 936.
[290] Rempel, G., P., Legzdins, H., Smith, G., Wilkinson, Inorg. Synth., 13, 1972, 90.
[291] Egan, D., Inorg. Chim. Acta, 359, 2006, 3976.
[292] Smith, J., L., Martin, Proc. Natl. Acad. Sci., 70, 1973, 1263.
[293] Erck, A., L., Rainen, J., Whileyman, J., Bear, Proc. Soc. Exp. Biol. Med., 145, 1974, 1278.
[294] Bear, J., H., Gray, L., Rainen, I., Chang, R., Howard, G., Serio, A., Kimball, Cancer Chemother. Rep. Part 1, 59, 1975, 611.
[295] Rainen, L., R., Howard, A., Kimball, J., Bear, Inorg. Chem. Trans., 14, 1979, 2752.
[296] Chifotides, H., K., Koshlap, L., Perez, K., Dunbar, J. Am. Chem. Soc., 125, 2003, 10703.
[297] Chifotides, H., K., Dunbar, Acc. Chem. Res., 38, 2005, 146.
[298] Clarke, M., F., Zhu, D., Frasca, Chem. Rev., 99, 1999, 2511.
[299] Jackson, B., J., Barton, J. Am. Chem. Soc., 119, 1997, 12986.
[300] Jackson, B., V., Alekseyev, J., Barton, Biochemistry, 38, 1999, 4655.
[301] Odom, D., C., Parker, J., Barton, Biochemistry, 38, 1999, 5155.
[302] Pruchnik, F., P., Jakimowicz, Z., Ciunik, J., Zakrzewska-czerwinska, A., Opolski, J., Wietrzyk, E., Wojdat, Inorg. Chim. Acta., 334, 2002, 59.
[303] Quezada, C. A., J. C., Garrison, M. J., Panzner, C. A., Tessier, W. J., Youngs, Organometallics, 23, 2004, 4846.
[304] Menon, E. L., R., Perera, M., Navarro, R. J., Kuhn, H., Morrison, Inorg. Chem., 43, 2004, 5373.
[305] Dwyer, F. P., E., Mayhew, E. M. F., Roe, A., Shulman, Brit. J. Cancer, 19, 1965, 195.
[306] Richens, D. T., The Chemistry of Aqua Ions, John-Wiley & Sons Ltd., Chichester, 1997.
[307] Allardyce, C. S., P. J., Dyson, Plat. Met. Rev., 45, 2001, 62.
[308] Clarke, M. J., Coord. Chem. Rev., 236, 2003, 209.
[309] Kostova, I., Curr. Med. Chem., 13, 2006, 1085.
[310] Alessio, E., G., Mestroni, A., Bergamo, G., Sava, Curr. Top. Med. Chem., 4, 2004, 1525.
[311] Clarke, M., Met. Ions Biol. Syst., 11, 1980, 231.

[312] Keppler, B., M., Henn, U., Juhl, M., Berger, R., Niebl, W., Balzer, Progr. Clin. Biochem. Med., 10, 1989, 41.
[313] Keller, H., B., Keppler, US Patent, 4843069, 1989.
[314] Alessio, E., G., Mestroni, A., Bergamo, G., Sava, Met. Ions Biol. Syst., 42, 2004, 323.
[315] Alessio, E., E., Iengo, S., Zorzet, A., Bergamo, M., Coluccia, A., Boccarelli, G., Sava, J. Inorg. Biochem., 79, 2000, 173.
[316] Sava, G., I., Capozzi, K., Clerici, G., Gagliardi, E., Alessio, G., Mestroni, Clin Exp. Metastas., 16, 1998, 371.
[317] Bergamo, A., G., Sava, Dalton Trans., 2007, 1267.
[318] Bergamo, A., G., Stocco, C., Casarsa, M., Cocchietto, E., Alessio, B., Serli, S., Zorzet, G., Sava, Int. J. Oncol., 24, 2004, 373.
[319] Vock, C. A., W. H., Ang, C., Scolaro, A. D., Phillips, L., Lagopoulos, L., Juillerat-jeanneret, G., Sava, R., Scopelliti, P. J., Dyson, J. Med. Chem., 50, 2007, 2166.
[320] Mestroni, G., E., Alessio, G., Sava, Int. Patent PCT C 07F 15/00, A61K 31/38, WO 98/00431, 1998.
[321] Morris, R., R., Aird, P., Murdoch, H., Chen, J., Cummings, N., Hughes, S., Parsons, A., Parkin, G., Boyd, D., Jodrell, P., Sadler, J. Med. Chem., 44, 2001, 3616.
[322] Hotze, A., M., Bacac, A., Velders, B., Jansen, H., Kooijman, A., Spek, J., Haasnoot, J., Reedijk, J. Med. Chem., 46, 2003, 1743.
[323] Vliet, P., S., Toekimin, J., Haasnoot, J., Reedijk, O., Novakova, O., Vrana, V., Brabec, Inorg. Chim. Acta, 231, 1995, 57.
[324] Kapitza, S., M., Pongratz, M., Jakupec, P., Heffeter, W., Berger, L., Lackinger, B., Keppler, B., Marian, J. Cancer Res., 131, 2005, 101.
[325] Kung, A., T., Pieper, R., Wissiack, E., Rosenberg, B. K., Keppler, J. Biol. Inorg. Chem., 6, 2001, 292.
[326] Piccioli, F., S., Sabatini, L., Messori, P., Orioli, C. G., Hartinger, B. K., Keppler, J. Inorg. Biochem., 98, 2004, 1135.
[327] Sava, G., S., Zorzet, T., Giraldi, G., Mestroni, G., Zassinovich, Eur. J. Cancer Clin. Oncol., 20, 1984, 841.
[328] Frasca, D. R., M. J., Clarke, J. Am. Chem. Soc., 121, 1999, 8523.
[329] Ang, W. H., P. J., Dyson, Eur. J. Inorg. Chem., 2006, 4003.
[330] Dougan, S. J., A., Habtemariam, S. E., McHale, S., Parsons, P. J., Sadler, Proc. Natl. Acad. Sci. U. S. A., 105, 2008, 11628.
[331] Habtemariam, A., M., Melchart, R., Fernandez, S., Parsons, I., Oswald, A., Parkin, F., Fabbiani, J. E., Davidson, A., Dawson, R. E., Aird, D. I., Jodrell, P. J., Sadler, J. Med. Chem., 49, 2006, 6858.
[332] Kostrhunová, H., J., Florian, O., Nováková, A. F. A., Peacock, P. J., Sadler, V., Brabec, J. Med. Chem., 51, 2008, 3635.
[333] Morris, R. E., R. E., Aird, P. D., Murdoch, H. M., Chen, J., Cummings, N. D., Hughes, S., Parsons, A., Parkin, G., Boyd, D. I., Jodrell, P. J., Sadler, J. Med. Chem., 44, 2001, 3616.
[334] Chvalová, K., J., Kaspárková, N., Farrell, V., Brabec, FEBS J., 273, 2006, 3467.
[335] Allardyce, C. S., P. J., Dyson, D. J., Ellis, S. L., Heath, Chem. Commun., 2001, 1396.
[336] Schibli, R., R., Netter, R., Waibel, U., Abram, A., Schbiger, Coord. Chem. Rev., 190–192, 1999, 901.
[337] Garcia, R., A., Paulo, A., Domingos, I., Santos, K., Ortner, R., Alberto, J. Am. Chem. Soc., 122, 2000, 11240.
[338] Wei, L., S., Banerjee, M., Levadala, J., Babich, J., Zubieta, Inorg. Chem. Comm., 6, 2003, 1099.
[339] Wang, W., B., Spingler, R., Alberto, Inorg. Chim. Acta, 355, 2003, 386.
[340] Bakir, M., Inorg. Chem., 332, 2002, 1.

[341] Cabeza, N., A., Garcia, M., Moreno-carretero, J., Martos, R., Exposito, J. Inorg. Biochem., 99, 2005, 1637.
[342] Ma, D., C., Che, F., Siu, M., Yang, K., Wong, Inorg. Chem., 46, 2007, 740.
[343] Baird, I., J. Med. Chem., 49, 2006, 5262.
[344] Lim, I., J. Am. Chem. Soc., 126, 2004, 10271.
[345] Postema, E. J., H. H., Boersma, Drugs Future, 29, 2004, 95.
[346] Wang, S.-J., T.-Y., Lo, B.-T., Hsieh, K.-Y., Cheng, W.-Y., Lin, L.-H., Shen, M.-J., Su, J. Radioanal. Nucl. Chem., 261, 2004, 189.
[347] Boschi, A., L., Uccelli, A., Duatti, P., Colamussi, C., Cittanti, A., Filice, A. H., Rose, A., Martindale, G., Claringbold, D., Kearney, H. J., Turner, M., Giganti, Nucl. Med. Commun., 25, 2004, 691.
[348] Luo, T.-Y., B.-T., Hsieh, S.-J., Wang, W.-Y., Lin, T.-W., Lee, L.-H., Shen, M.-J., Su, Nucl. Med. Biol., 31, 2004, 671.
[349] Faintuch, B. L., S., Faintuch, E., Muramoto, J. Radioanal. Nucl. Chem., 260, 2004, 363.
[350] Fernandes, C., T., Kniess, L., Gano, S., Seifert, I., Santos, Nucl. Med. Biol., 31, 2004, 785.
[351] He, J., C., Liu, J.-L., Vanderheyden, G., Liu, S., Dou, M., Rusckowski, D. J., Hnatowich, Nucl. Med. Commun., 25, 2004, 731.
[352] Gorshkov, N. I., R., Schibli, A. P., Schubiger, A. A., Lumpov, A. E., Miroslavov, D. N., Suglobov, J. Organomet. Chem., 689, 2004, 4757.
[353] Ogawa, K., T., Mukai, Y., Arano, H., Hanaoka, K., Hashimoto, H., Nishimura, H., Saji, J. Labelled Compd. Radiopharm., 47, 2004, 753.
[354] Lam, M., J., De Klerk, P. P., Van Rijk, Eur. J. Nucl. Med. Mol. Imaging, 31, 2004, S162.
[355] Jiang, S.-B., S.-Z., Luo, H.-F., Deng, W.-Z., Bin, W.-J., Wang, H.-Y., Wei, G.-P., Liu, Nucl. Sci. Techn., 15, 2004, 39.
[356] Frier, M., Mini-Rev. Med. Chem., 4, 2004, 61.
[357] Top, S., A., Vessieres, P., Pigeon, M.-N., Rager, M., Huche, E., Salomon, C., Cabestaing, J., Vaissermann, G., Jaouen, ChemBioChem, 5, 2004, 1104.
[358] Lansdown, A., J. Wound Care., 11, 2002, 173.
[359] Drake, P., Ann. Occup. Hyg., 49, 2005, 575.
[360] Boswald, M., K., Mende, W., Bernschneider, S., Bonakdar, H., Ruder, H., Kissler, Infection., 27, 1999, 38.
[361] Brutel, A., K., Dossche, D., Binbaum, R., Hacker, J. Heart Valve Disc., 9, 2000, 123.
[362] Hostynyk, J., R., Hinz, C., Lorence, M., Price, R., Guy, Tox., 1993, 171.
[363] Berners-price, S., R., Johnson, A., Giovenella, L., Faucette, C., Mirabelli, P., Sadler, J. Inorg. Biochem., 33, 1988, 285.
[364] Liu, J., P., Galetis, A., Farr, L., Maharaj, H., Samarasinha, A., McGechan, B., Baguley, R., Bowen, S., Berners-price, M., McKeage, J. Inorg. Biochem., 102, 2008, 303.
[365] Zhu, H., X., Zhang, X., Liu, X., Wang, G., Liu, A., Usman, H., Fun, Inorg. Chem. Comm., 6, 2003, 1113.
[366] Liu, Z., Z., You, H., Zhu, M., Tan, Inorg. Chem. Comm., 7, 2004, 1292.
[367] Zachariadis, P., S., Hadjikakou, N., Hadjiliadis, S., Skoulika, A., Michaleides, J., Balzarini, E., De Clerq, Eur. J. Inorg. Chem., 2004, 1420.
[368] Gottscaldt, M., A., Pfeifer, D., Koth, H., Gorls, H., Dahse, U., Mollmann, M., Obata, S., Yano, Tetrahedron, 62, 2006, 11073.
[369] Thati, B., A., Noble, B., Creaven, M., Walsh, M., McCann, K., Kavanagh, M., Devereux, D., Egan, Cancer Lett., 248, 2007, 321.
[370] Thati, B., A., Noble, B., Creaven, M., Walsh, M., McCann, K., Kavanagh, M., Devereux, D., Egan, Cancer Lett., 250, 2007, 128.

[371] Egan, D., Inorg. Chim. Acta., 359, 2006, 3976.
[372] Hidalgo, E., C., Domiguez, Tox. Lett., 98, 1998, 169.
[373] Smith, J., L., Martin, Proc. Natl. Acad. Sci, 70, 1973, 1263.
[374] Peyrone, M., Ann. Chemie Pharm., 1845, 51.
[375] Rosenberg, B. L., T., Krigas, Nature, 205, 1965, 698.
[376] Bertini, I., Bioinorganic Chemistry: Inorganic Elements in the Chemistry of Life-An Introduction and Guide In Bioinorganic Chemistry, Ed., John Wiley and Sons, 1994.
[377] Weiss, R. B., M. C., Christian, Drugs, 46, 1993, 360.
[378] Lebwohl, D., R., Canetta, Eur. J. Cancer, 34, 1998, 1522.
[379] Xu, H. D., S. M., Choi, C. S., An, Y. D., Min, K. C., Kim, K. J., Kim, C. H., Choi, Biochem. Biophys. Res. Commun., 328, 2005, 618.
[380] Lee, J. W., J. K., Park, S. H., Lee, S. Y., Kim, H. J., Kuh, Anti-Cancer Drugs, 17, 2006, 377.
[381] Park, Y. M., S., Bae, W., Chung, B., Byun, J., Choi, S., Cha, K., Chung, H., Sun, D., Park, B., Choi, Hepatology, 34, 2001, 673A.
[382] Oh, J. M., S., Gashil, Pharmacotherapy, 22, 2002, 1357.
[383] Kang, J. H., H. J., Kuh, J. H., Lee, J. Y., Shin, K. S., Lee, J. A., Jung, D. Y., Chang, J. Clin. Oncol., 23, 2005, 512S.
[384] Kim, H. T., D. H., Lee, K. H., Jung, J. Y., Han, I. J., Jang, H. S., Lim, S. H., Park, Y. S., Park, J., Ro, J. S., Lee, Eur. J. Cancer Suppl., 3, 2005, 422.
[385] Jamieson, E. R., S. J., Lippard, Chem. Rev., 99, 1999, 2467.
[386] Dwyer, P. J., S. W., Johnson, 30 Years of Cisplatin, Chemistry & Biochemistry of a Leading Anticancer Drug, Lippert, B., Ed., Wiley-VCH, Weinheim, Verlag Zurich, 1999.
[387] Chu, G., J. Biol. Chem., 269, 1994, 787.
[388] Pérez, R. P., Eur. J. Cancer, 34, 1998, 1535.
[389] Kelland, L. R., Drugs, 59, 2000, 1.
[390] Cleare, M., Plat. Met. Rev., 17, 1973, 2.
[391] Cleare, M., Bioinorg. Chem., 2, 1973, 187.
[392] Rixe, O., W., Ortuzar, M., Alvarez, R., Parker, E., Reed, K., Paull, T., Fojo, Biochem. Pharmacol., 52, 1996, 1855.
[393] Gietema, J. A., E. G. E., Devries, D. T., Sleijfer, P. H. B., Willemse, H. J., Guchelaar, D. R. A., Uges, P., Aulenbacher, R., Voegeli, N. H., Mulder, Brit. J. Cancer, 67, 1993, 396.
[394] Markman, M., L. C., Demarco, M., Birkhofer, D., Budman, T., Hakes, B., Reichman, S., Rubin, W., Jones, R., Barakat, J., Curtin, J. L., Lewis, L., Almadrones, A., Hoffman, R., Rastogi, W., Hoskins, J. Cancer Res. Clin., 119, 1993, 234.
[395] Wheate, N. J., C. R., Brodie, J. G., Collins, S., Kemp, J. R., Aldrich-wright, Mini-Rev. Med. Chem., 7, 2007, 627.
[396] Kaspárková, J., M., Vojtisková, G., Natile, V., Brabec, Chem. Eur. J., 14, 2008, 1330.
[397] Malina, J., O., Nováková, M., Vojtisková, G., Natile, V., Brabec, Biophys. J., 93, 2007, 3950.
[398] Spingler, B., D. A., Whittington, S. J., Lippard, Inorg. Chem., 40, 2001, 5596.
[399] Hambley, T. W., Coord. Chem. Rev., 166, 1997, 181.
[400] Noji, M., K., Okamoto, Y., Kidani, J. Med. Chem., 24, 1981, 508.
[401] Dragovich, T., D., Mendelson, S., Kurtin, K., Richardson, D., Von Hoff, A., Hoos, Cancer Chemoth. Pharm., 58, 2006, 759.
[402] MacLean, D. S., A., Khokhar, R., Pérez-soler, Cancer Biother. Radiopharm., 15, 2000, 253.
[403] Verschraegen, C. F., S., Kumagai, R., Davidson, B., Feig, P., Mansfield, S. J., Lee, D. S., Maclean, W., Hu, A. R., Khokhar, Z. H., Siddik, J. Cancer Res. Clin., 129, 2003, 549.
[404] Lu, C., R., Pérez-soler, B., Piperdi, G. L., Walsh, S. G., Swisher, W. R., Smythe, H. J., Shin, J. Y., Ro, L., Feng, M., Truong, D. M., Shin, J. Clin. Oncol., 23, 2005, 3495.
[405] Saito, T., Y., Manabe, H., Saito, J. Oto-Rhino-Laryngol. Relat. Spec., 57, 1995, 250.

[406] Shi, Y. C., S., Inoue, R., Shinozaki, K., Fukue, T., Kougo, Jpn. J. Cancer Res., 89, 1998, 757.
[407] Munk, V. P., C. I., Diakos, L. T., Ellis, R. R., Fenton, B. A., Messerle, T. W., Hambley, Inorg. Chem., 42, 2003, 3582.
[408] Kelland, L. R., S. Y., Sharp, C. F., O'neill, F. I., Raynaud, P. J., Beale, I. R., Judson, J. Inorg. Biochem., 77, 1999, 111.
[409] Holford, J., F., Raynaud, B. A., Murrer, K., Grimaldi, J. A., Hartley, M., Abrams, L. R., Kelland, Anti-Cancer Drug Des., 13, 1998, 1.
[410] Raynaud, F. I., F., Boxall, P. M., Goddard, M., Valenti, M., Jones, B. A., Murrer, M., Abrams, L. R., Kelland, Clin. Cancer Res., 3, 1997, 2063.
[411] Kalinowska, U., J., Ochocki, K., Matlawska-wasowska, Coord. Chem. Rev., 252, 2008, 1328.
[412] Bierbach, U., Y., Qu, T. W., Hambley, J., Peroutka, H. L., Nguyen, M., Doedee, N., Farrell, Inorg. Chem., 38, 1999, 3535.
[413] Nováková, O., J., Kaspárková, J., Malina, G., Natile, V., Brabec, Nucleic Acids Res., 31, 2003, 6450.
[414] Boccarelli, A., D., Giordano, G., Natile, M., Coluccia, Biochem. Pharmacol., 72, 2006, 280.
[415] Vinje, J., F. P., Intini, G., Natile, E., Sletten, Chem. Eur. J., 10, 2004, 3569.
[416] Pantoja, E., A., Alvarez-valdes, J. M., Pérez, C., Navarro-ranninger, J., Reedijk, Inorg. Chim. Acta, 339, 2002, 525.
[417] Pantoja, E., A., Gallipoli, S., Van Zutphen, S., Komeda, D., Reddy, D., Jaganyi, M., Lutz, D. M., Tooke, A. L., Spek, C., Navarro-ranninger, J., Reedijk, J. Inorg. Biochem., 100, 2006, 1955.
[418] Pantoja, E., A., Gallipoli, S., Van Zutphen, D. M., Tooke, A. L., Spek, C., Navarro-ranninger, J., Reedijk, Inorg. Chim. Acta, 359, 2006, 4335.
[419] Volckova, E., E., Weaver, R. N., Bose, Eur. J. Med. Chem., 43, 2008, 1081.
[420] McFadyen, W. D., L., Wakelin, I. A. G., Roos, V. A., Leopold, J. Med. Chem., 28, 1985, 1113.
[421] Lowe, G., T., Vilaivan, J. Chem. Soc., Perkin Trans., 1, 1996, 1499.
[422] Eryazici, I., C. N., Moorefield, G. R., Newkome, Chem. Rev., 108, 2008, 1834.
[423] Brabec, V., J., Kaspárková, O., Vrána, O., Nováková, J. W., Cox, Y., Qu, N., Farrell, Biochemistry, 38, 1999, 6781.
[424] Farrell, N., J. Inorg. Biochem., 74, 1999, 23.
[425] Perego, P., C., Caserini, L., Gatti, N., Carenini, S., Romanelli, R., Supino, D., Colangelo, I., Viano, R., Leone, G., Pezzoni, C., Manzotti, F., Zunino, Mol. Pharmacol., 55, 1999, 1108.
[426] Perego, P., C., Caserini, L., Gatti, N., Carenini, S., Romanelli, R., Supino, D., Colangelo, I., Viano, R., Leone, S., Spinelli, C., Manzotti, N., Farrell, F., Zunino, Mol. Pharmacol., 55, 1999, 528.
[427] Perego, P., L., Gatti, C., Caserini, R., Supino, D., Colangelo, R., Leone, S., Spinelli, N., Farrell, F., Zunino, J. Inorg. Biochem., 77, 1999, 59.
[428] Pratesi, G., P., Perego, D., Polizzi, S. C., Righetti, R., Supino, C., Caserini, C., Manzotti, F. C., Giuliani, G., Pezzoni, S., Spinelli, N., Farrell, F., Zunino, Br. J. Cancer, 80, 1999, 1912.
[429] Rauter, H., R., DiDomenico, E., Menta, Y., Qu, N., Farrell, Inorg. Chem., 36, 1997, 3919.
[430] Roberts, J. D., J., Peroutka, N., Farrell, J. Inorg. Biochem., 77, 1999, 51.
[431] Wheate, N. J., J. G., Collins, Coord. Chem. Rev., 241, 2003, 133.
[432] Farrell, N. P., S. G., Dealmeida, K. A., Skov, J. Am. Chem. Soc., 110, 1988, 5018.
[433] Farrell, N., T. G., Appleton, Y., Qu, J. D., Roberts, A., Fontes, K. A., Skov, P., Wu, Y., Zou, Biochemistry, 34, 1995, 15480.
[434] Farrell, N., Y., Qu, M. P., Hacker, J. Med. Chem., 33, 1990, 2179.
[435] Kloster, M., H., Kostrhunova, R., Zaludova, J., Malina, J., Kasparkova, V., Brabec, N., Farrell, Biochemistry, 43, 2004, 7776.
[436] Hills, C. A., L. R., Kelland, J., Siracky, A., Wilson, K. R., Harrap, Br. J. Cancer, 59, 1989, 527.
[437] Pérez, J. M., M. A., Fuertes, C., Alonso, C., Navarro, Crit. Rev. Oncol. Hematol., 35, 2000, 109.
[438] Kaspárková, J., J., Zehnulová, N., Farrell, V., Brabec, J. Biol. Chem., 277, 2002, 48076.

[439] Lippert, B., Cisplatin, Chemistry and Biochemistry of a leading Antitumor Drug, Wiley-VCH, Zurich, 1999.
[440] Jodrell, D. I., T., Evans, W., Steward, D., Cameron, J., Prendiville, C., Aschele, C., Noberasco, M., Lind, J., Carmichael, N., Dobbs, B., Gatti, F., De Braud, Eur. J. Cancer, 40, 2004, 872.
[441] Billecke, C., S., Finniss, L., Tahash, C., Miller, T., Mikkelsen, N. P., Farrell, O., Boegler, Neuro-Oncology, 8, 2006, 215.
[442] Kabolizadeh, P., J., Ryan, N., Farrell, Biochem. Pharmacol., 73, 2007, 1270.
[443] Hague, D. N., A. D., Moreton, J. Chem. Soc., Perkin Trans., 2, 1994, 265.
[444] Feuerstein, B. G., L. D., Williams, H. S., Basu, L. J., Marton, J. Cell. Biochem., 46, 1991, 37.
[445] Navarro-ranninger, C., P. A., Ochoa, J. M., Pérez, V. M., Gonzalez, J. R., Masaguer, C., Alonso, J. Inorg. Biochem., 53, 1994, 177.
[446] Behe, M., G., Felsenfeld, Proc. Natl. Acad. Sci. U.S.A. Biol. Sci., 78, 1981, 1619.
[447] Jain, S., G., Zon, M., Sundaralingam, Biochemistry, 28, 1989, 2360.
[448] Hegmans, A., Y., Qu, L. R., Kelland, J. D., Roberts, N., Farrell, Inorg. Chem., 40, 2001, 6108.
[449] Hegmans, A., S. J., Berners-price, M. S., Davies, D. S., Thomas, A. S., Humphreys, N., Farrell, J. Am. Chem. Soc., 126, 2004, 2166.
[450] Kostova, I., Rec. Pat. Rev. Anti-Canc. Drug Disc., 1, 2006, 1.

Index

https://doi.org/10.1515/9781501516115-017